Welding - Materials, Fabrication Processes, and Industry 5.0

Edited by Sanjeev Kumar

Published in London, United Kingdom

Welding - Materials, Fabrication Processes, and Industry 5.0
http://dx.doi.org/10.5772/intechopen.1000354
Edited by Sanjeev Kumar

Contributors
Sanjeev Kumar, Margareth Nascimento de Souza Lira, Sadek Crisostomo Absi Alfaro, Holger Seidlitz, Lars Ulke-Winter, Lucas Ost, Felix Kuke, Jiangchao Wang, Bitao Liu, Zhangjing Bao, Wencheng Jiang, Zichao Zhuo, Langxiong Gan, Yaqing Shu, Jairo José Muñoz Chávez, Gerardo Antonio Idrobo Pizo, Sadek Crisostomo Absi Alfaro, Gilbert Tukahirwa, Catherine Wandera

First published in London, United Kingdom, 2024 by IntechOpen
IntechOpen is the global imprint of INTECHOPEN LIMITED, registered in England and Wales, registration number: 11086078, 5 Princes Gate Court, London, SW7 2QJ, United Kingdom

British Library Cataloguing-in-Publication Data
A catalogue record for this book is available from the British Library

Additional hard and PDF copies can be obtained from orders@intechopen.com

Welding - Materials, Fabrication Processes, and Industry 5.0
Edited by Sanjeev Kumar
p. cm.
Print ISBN 978-1-83769-871-4
Online ISBN 978-1-83769-870-7
eBook (PDF) ISBN 978-1-83769-872-1

For EU product safety concerns:
IN TECH d.o.o., Prolaz Marije Krucifikse Kozulić 3, 51000 Rijeka, Croatia,
info@intechopen.com or visit our website at intechopen.com.

Meet the editor

Dr. Sanjeev Kumar is an assistant professor in the Department of Mechanical Engineering, National Institute of Technology Raipur, India. Prior to this position, he was a research associate at the Christian Doppler Laboratory for Interfaces and Precipitation Engineering (CDL-IPE), Vienna University of Technology, Austria, for almost four years while working as a postdoctoral researcher on the FWF (Austrian Science Fund) project at the Institute of Materials Science, Joining and Forming (IMAT), Graz University of Technology, Austria. He obtained a BTech in Mechanical Engineering from Uttar Pradesh Technical University (UPTU), Lucknow, India, and an MTech and Ph.D. in Metallurgy from the Indian Institute of Technology Roorkee, India. His primary research interests are welding, thermomechanical processing, material design, physical metallurgy, and Industry 5.0. Dr. Kumar has published two books and more than forty-five research papers in reputed international journals and conferences.

Contents

Preface

In the present era, the demand for new products with attractive designs and highly configured features is increasing. Innovative products are being developed using the latest technology and appropriate material grades. As such, both the physical and chemical properties of the materials used are important. It is well known that all required physical and mechanical properties cannot be achieved in a single grade of material, therefore it is exigent to determine the technology by which similar and dissimilar kinds of materials can be joined without any leakages or defects. In 3000 BC, archaeologists demonstrated a joint between two pieces by hammering dating under heat, a process now known as welding. It is common to define welding as "a materials joining process in which two (or more) parts coalesced at their contacting surfaces by the suitable application of heat and/or pressure". The mechanical properties of the joining materials deteriorate due to high heat and pressure but are improved by a secondary heat treatment process or by controlling appropriate welding parameters. Hence, both time and cost for product development increase due to having to use several types of manufacturing processes, which is not preferable for present-day industries. However, the improvement in manufacturing facilities and high strength-to-weight ratio materials are key to developing innovative products and launching them in the market with minimum time and at reasonable cost.

This book, *Welding – Materials, Fabrication Processes, and Industry 5.0*, is a collection of examples illustrating research in welding technology with special emphasis on the latest technology. Chapter 1 provides information on the latest trends in technology. Chapter 2 discusses state-space modeling of weld bead geometry in the gas metal arc-directed energy deposition process applied to wire and arc additive manufacturing and welding processes. Chapter 3 describes appropriate material and load path joining techniques for FRP/metal hybrid structures. Chapter 4 presents information on the ongoing energy savings and emission reductions during welding fabrication processing. Chapter 5 develops a methodology for the deposition process in gas metal arc welding. Chapter 6 discusses the influence of process parameters in gas metal arc welding of carbon steels.

This book is a useful resource for engineering students as well as academics, researchers, professionals in welding technology and related industries, and mechanical, manufacturing, industrial, and materials engineers. We hope this book will inspire readers to undertake research in welding technology. The editor acknowledges IntechOpen for the opportunity to edit this book and for their enthusiastic and professional support. Finally, I would like to thank all the chapter authors for their excellent contributions.

Dr. Sanjeev Kumar
Department of Mechanical Engineering,
National Institute of Technology Raipur,
Chhattisgarh, India

Section 1

Introduction and Latest Upgrades in Technology

Chapter 1

Introductory Chapter: Welding in the Era of Industry 5.0

Sanjeev Kumar

1. Introduction

The outstanding advancement of manufacturing and industrial processes may be seen in the journey from Industry 1.0 to Industry 5.0 [1–5], as shown in **Figure 1**. With the emergence of Industry 1.0 in the late 1700s, agricultural economies saw a dramatic shift to mechanized industry based on steam and water power. Revolutionary innovations like steam engine and mechanized looms were introduced during this time. During the late nineteenth and early twentieth centuries, Industry 2.0 saw the rise of electrification and mass manufacturing. Developments like electricity and the assembly line made it easier to produce standardized items on a large scale in an efficient manner [7]. The Digital Revolution, or Industry 3.0, began in the late twentieth century with the advent of computer technology, early robots, and automation. Information technology and computer-controlled manufacturing saw a paradigm change as a result [8].

The early twenty-first century marked the advent of Industry 4.0, the Fourth Industrial Revolution. It centered on automation, data exchange, Internet of Things (IoT), and cyber-physical systems [9–11]. This era facilitated the development of smart factories, predictive maintenance, and the concept of digital twins. In the emerging era of Industry 5.0, the focus is on the collaboration of humans and machines [7]. This phase introduces collaborative robots (COBOTS) working in collaboration with humans. The concept behind this collaboration is to combine human creativity and decision-making skills with advanced technologies to develop a more innovative, efficient, and adaptive industrial environment [12]. COBOTS assist humans in tasks that require dexterity, precision, and rapid adaptation [13]. Even in complicated, nonroutine circumstances, this collaboration helps in quick problem-solving and improving productivity with enhanced product quality. As a means of bridging the gap between human expertise and technological capabilities, Industry 5.0 envisions manufacturing as a future where innovation and adaptation are critical.

2. A comparison between current (Industry 4.0) and proposed (Industry 5.0) industrial revolution

The comparison between current (Industry 4.0) and proposed (Industry 5.0) industrial revolution based on various aspects is tabled in **Table 1** [2, 18]:

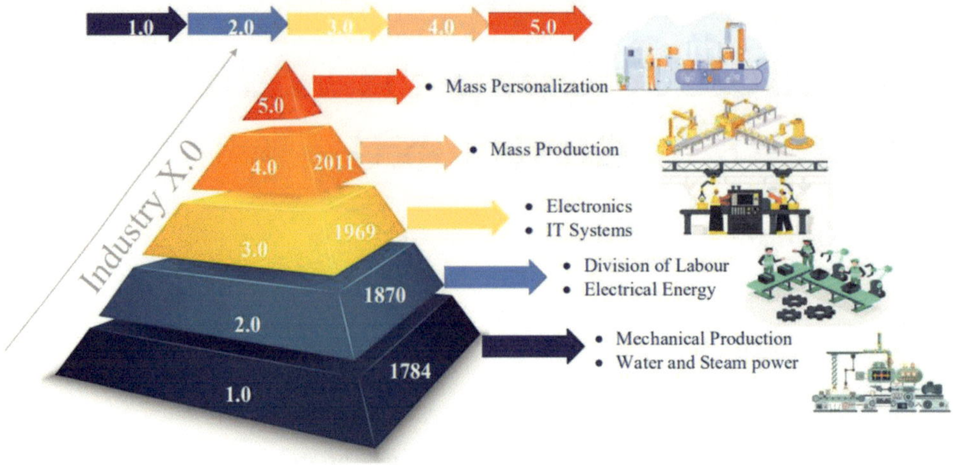

Figure 1.
Evolution of industry, from Industry 1.0 to Industry 5.0 [6].

Aspect	Current industrial revolution, Industry 4.0	Proposed industrial revolution, Industry 5.0
Emphasis	Automation and digitization	Human–machine collaboration [10]
Key technology	IoT, big data, AI, and automation [2]	Collaborative robotics [10], AI decision support, and hyper-automation [14]
Human involvement	Humans in a supervisory role	Humans actively collaborate with machines
Role of automation	Extensive automation of processes	Automation complements human capabilities
Customization	Focus on customization and batch size	Individualization of products and "mass customization" [15]
Complexity and problem-solving	Limited human intervention in complex tasks	Humans actively address complex problems
Real-time decision support	Limited, often machine-driven	AI provides real-time decision support [16]
Interoperability	Emphasized for machine-to-machine	Collaboration between human and machine
Data analytics	Big data and analytics for process optimization	Data analytics for insights and decision support
Transition	Transition from manual to automated processes	Transition from full automation to collaboration [17]

Table 1.
Status of technology development in Industry 4.0 and proposed Industry 5.0.

3. The increasing role of AI and ML in manufacturing

The technologies of Artificial Intelligence (AI) and Machine Learning (ML) are revolutionary, and their goal is to give machines the capacity to think like humans do, that is, to learn, reason, and make judgments. The detailed classification of Artificial Intelligence is shown in **Figure 2**. AI is the larger area that includes a variety of approaches that allow machines to emulate intelligent behavior. ML, a subset of AI, is

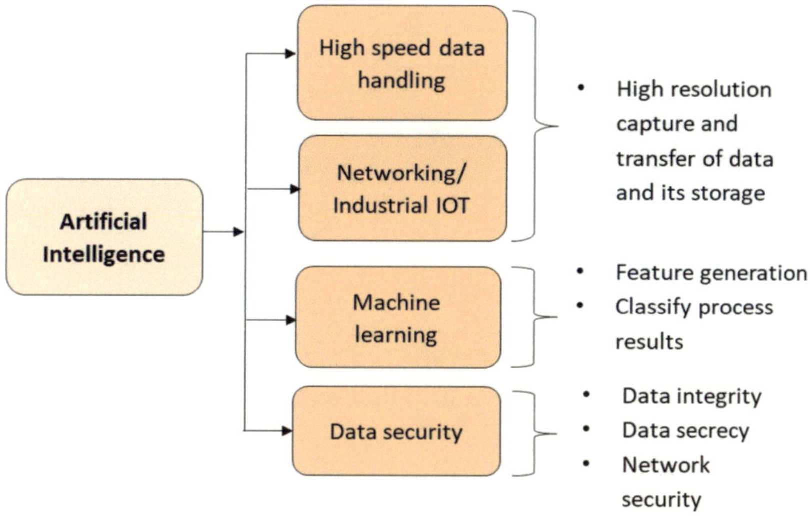

Figure 2.
Artificial intelligence.

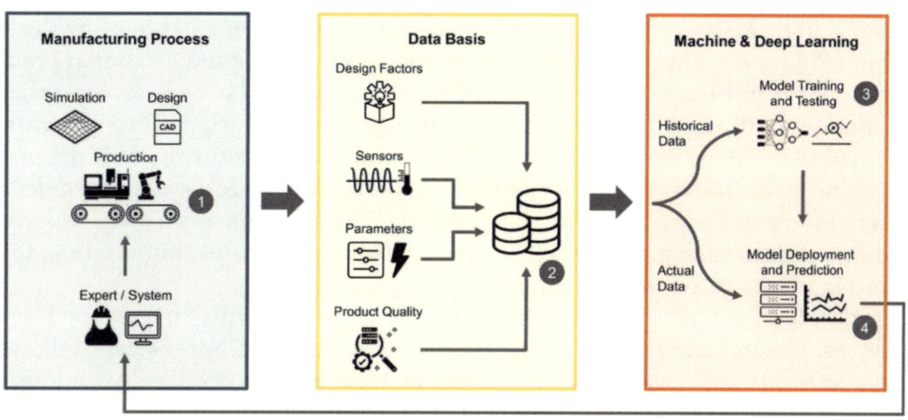

Figure 3.
Data-driven manufacturing. (1) particular manufacturing process, (2) relevant process data collection, (3) using data as a basis for training an ML model, and (4) use of trained model, to perform quality estimations for decision support [19].

the study of creating algorithms that let computers learn from and make predictions or judgments based on data [11]. The information of different modes such as physical, simulation or data basis, response of machines can be seen in **Figure 3** in detail.

The integration of AI and ML into manufacturing is a transformative phenomenon that is revolutionizing the industry on multiple fronts. Complex product assembly is made possible by AI-driven robotics and automation, with excellent accuracy and efficiency [20]. AI-powered quality control systems enable real-time defect detection by analyzing sensor data and images, which allows for immediate corrective action and guarantees high-quality output. ML improves manufacturing processes by evaluating large datasets to find patterns and trends [21], and AI improves predictive maintenance, which reduces the downtime and maintenance expenses [3]. With the help of these technologies, manufacturing systems will become more responsive, flexible, and agile.

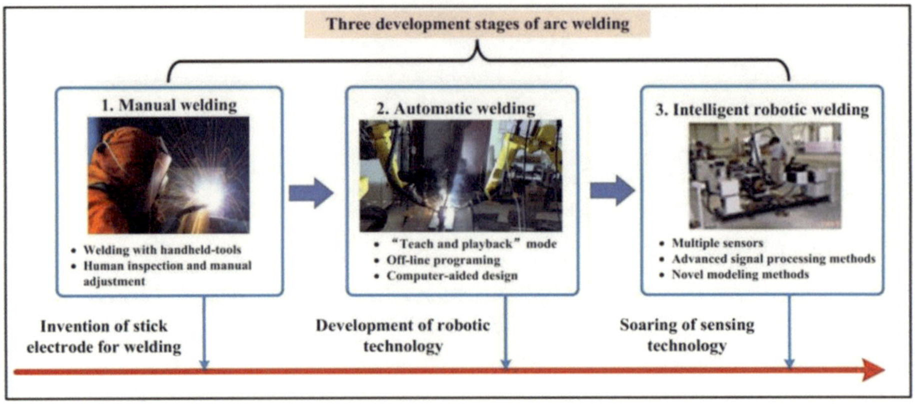

Figure 4.
Different development stages of arc welding [22].

4. The importance of welding in manufacturing

Welding is an important manufacturing process that creates durable joins in materials, typically metals or thermoplastics, by melting their edges and blending them together (**Figure 4**). The purpose of joint is to join two similar and dissimilar grades of materials for fulfilling the desired mechanical and physical properties in application. Sometimes the resulting weld is often stronger than the original components due performed thermal cycle and filler materials. Beyond manufacturing, welding plays a vital role in repair and maintenance, especially in sectors like aerospace and defense, to ensure safety and precision. Welding serves as the backbone of various industries, including manufacturing, construction, automotive, energy, and shipbuilding, for the fabrication of structures and components [23].

The development of welding spans centuries, beginning with early manual techniques. The Industrial Revolution introduced mechanized approaches, followed by twentieth-century advancements. The transformative leap occurred with robotic welding, offering unparalleled accuracy, precision, and safety. Integration of AI and ML in recent years has further elevated robotic welding, enabling adaptive control, real-time optimization, defect detection, and predictive maintenance.

The evolving welding industry demands innovation to meet modern industrial demands. Advanced techniques are crucial for enhancing production and weld quality along with worker's safety. Also, enhancing sustainability involves the use of resource optimization and energy-efficient techniques. By overcoming these challenges and taking advantage of these opportunities, welding productivity, quality, and sustainability are all increased.

5. Robotic welding and automation

Automation and robotic welding both refer to the automated and controlled use of robotic devices for welding processes as shown in **Figure 5**. These systems employ robotic arms or manipulators that are equipped with welding instruments to perform welding operations with enhanced output, better quality, and less costs. This technique is used in a variety of industries, such as manufacturing, construction,

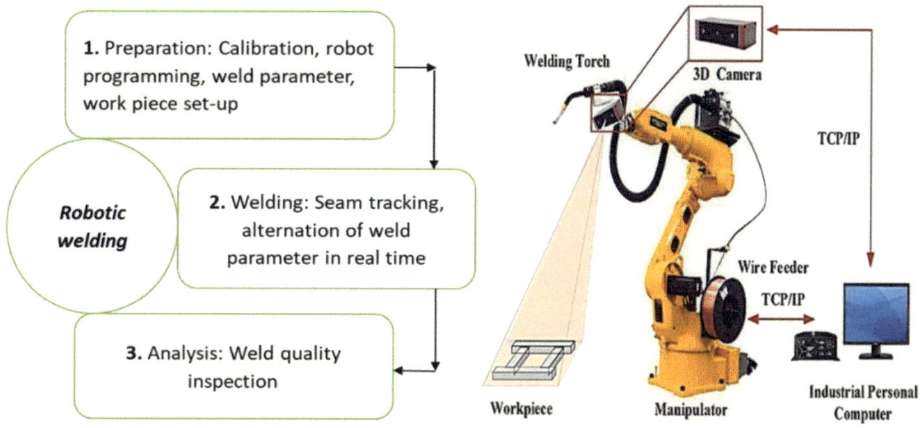

Figure 5.
(a) Steps in robotic welding [24] and (b) the robotic welding system [25].

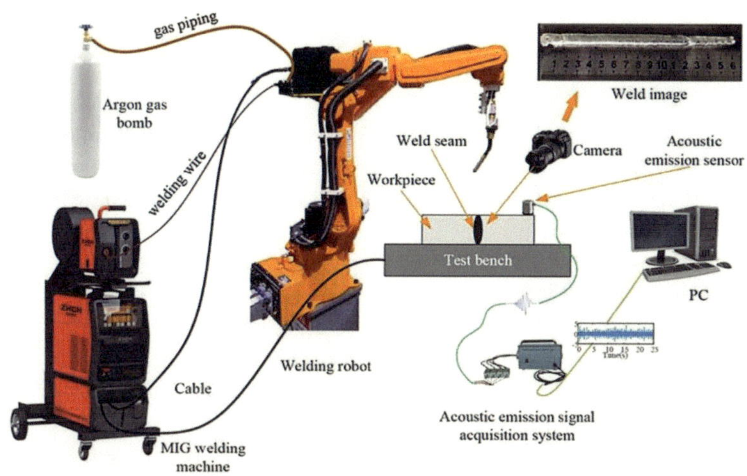

Figure 6.
Robotic MIG welding [27].

and automotive, where robotic systems can effectively manage complex or repetitive welding processes [24–26].

Robotic welding and automation have revolutionized the welding industry by enhancing not only accuracy and precision but also efficiency and productivity. The choice of robotic welding type depends on material types, joint configurations, and usages. Arc welding robots create robust joints through electric arcs on metal surfaces and are commonly used in automotive and construction industries. Spot welding robots create joints on metal sheets at specific points. Plasma, laser, and friction welding robots provide high-precision needs. MIG (**Figure 6**) and TIG welding robots offer high-quality welds, with MIG suitable for automotive applications and TIG for precision in aerospace.

Benefits of robotic welding and automation:

1. Robotic welding ensures precision and high-quality welds consistently.

2. Robots work at a steady and efficient pace, which leads to faster production rates.

3. Complex and intricate welding tasks, which are challenging for manual welding, can be easily achieved with robots.

4. Modern robotic welding systems offer easy reprogramming, adaptability to diverse tasks, and scalability to efficiently meet changing production requirements.

5. Robots ensure workplace safety by handling tasks in extreme environments, such as welding fumes, extreme heat, and so on.

6. How AI and ML can revolutionize welding processes

Welding procedures become more efficient and cost-effective when AI and ML with automation and robotics are used. AI and ML systems help in improving welding operations' safety, dependability, and compliance with industry requirements in different sectors [28–31], as shown in **Figures 7 and 8**. There are following ways in which AI and ML provide consistent weld quality:

1. *Real-time monitoring*: real-time welding process monitoring is made possible by AI and ML. Any change from the standard initiates quick corrections. This ensures that regardless of changing circumstances, welds must adhere to the intended requirements

2. *Quality control and defect detection*: throughout the welding process, continuous collection of data through real-time sensor data, photos, and videos allows AI and ML systems to learn and improve over time. This feedback loop also allows for quick remedial action by identifying flaws like porosity, fractures, or partial penetration. Overall weld quality is improved by early identification and correction.

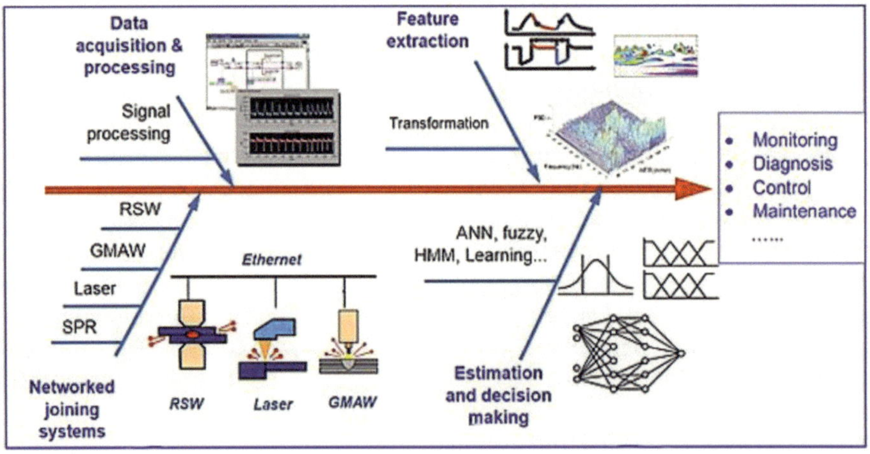

Figure 7.
Flowchart of process monitoring and control in welding [31].

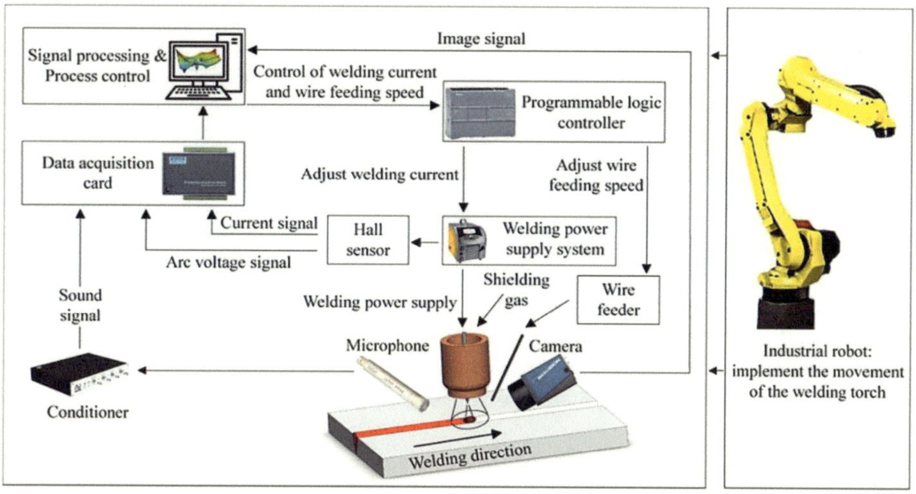

Figure 8.
System architecture of TIG welding work cell [22].

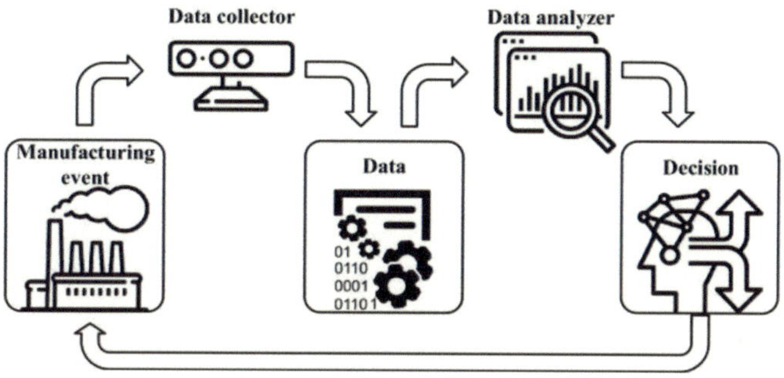

Figure 9.
Data-driven manufacturing [32].

3. *Data-driven insights and decision-making*: AI and ML evaluate enormous volumes of historical welding data, to find trends and connections between welding parameters and weld quality. With the use of these technologies, data-driven decision support is possible, which helps in making wise decisions during intricate and dynamic welding situations (**Figure 9**).

4. *Adaptive control*: machine learning algorithms provide consistent weld quality in a variety of circumstances by adapting to different materials, joint geometry, and environmental factors. Robots can also be reprogrammed with ML algorithm to enhance the response, according to change in workpiece alignment and other settings.

5. *Welding process optimization*: welding parameters like voltage, current, wire feed speed, and gas flow can be optimized using ML models. These algorithms ensure the right parameters to be used at the right level for every welding process by adapting to differences in materials and ambient circumstances.

6. *Predictive maintenance*: by analyzing the sensor data of welding machines or components, AI can predict when welding equipment may require maintenance or calibration. It prevents quality issues due to equipment malfunction, prevents unexpected breakdowns, and lowers maintenance costs.

7. *Welding simulation and training*: ML techniques are utilized to generate realistic welding simulations for teaching purposes. These are especially useful for conducting tests on various welding settings in the absence of tangible prototypes and training welders.

8. *Resource efficiency*: AI and ML can help in optimizing and maximizing the utilization of resources, including energy and other consumables to make welding process more economical, sustainable, and environmentally friendly.

9. *Enhanced worker safety*: AI and ML help in improving the worker safety and reducing accidents and exposure to danger by tracking working circumstances and giving real-time feedback and alerts.

7. Benefits and challenges of AI and ML in welding

Benefits:

1. AI-driven welding systems detect defects in real-time, that helps in minimizing at the moment.

2. ML algorithms optimize welding parameters, scheduling, and resource usage that improve production time and reduce waste.

3. AI predicts equipment issues based on data and suggests when to provide maintenance, which helps in reducing downtime.

4. AI enhances safety by minimizing human exposure to hazards.

Challenges:

1. AI and ML heavily rely on quality data. Thus, ensuring accurate, comprehensive data collection and data preprocessing is a very important task.

2. Implementing AI and ML can be complex. So, training and expert support for continuous integration is essential.

3. Understanding AI decisions are complicated; thus, transparent AI models must be developed.

4. Reliable sensors and equipment are required for continuous data collection, which require large investment.

5. AI systems must adhere to regulations for compliance with industry standards.

Author details

Sanjeev Kumar
Department of Mechanical Engineering, National Institute of Technology Raipur, Chhattisgarh, India

*Address all correspondence to: skumar.me@nitrr.ac.in

References

[1] Vinitha K, Ambrose Prabhu R, Bhaskar R, Hariharan R. Review on industrial mathematics and materials at industry 1.0 to industry 4.0. Materials Today: Proceedings. 2020;**33**:3956-3960. DOI: 10.1016/j.matpr.2020.06.331

[2] Akundi A, Euresti D, Luna S, Ankobiah W, Lopes A, Edinbarough I. State of industry 5.0—Analysis and identification of current research trends. Applied System Innovation. 2022;**5**(1):1-14. DOI: 10.3390/asi5010027

[3] Khan M, Haleem A, Javaid M. Changes and improvements in industry 5.0: A strategic approach to overcome the challenges of industry 4.0. Green Technologies and Sustainability. 2023;**1**(2):100020. DOI: 10.1016/j.grets.2023.100020

[4] Mourtzis D, Angelopoulos J, Panopoulos N. A literature review of the challenges and opportunities of the transition from industry 4.0 to society 5.0. Energies. 2022;**15**(17). DOI: 10.3390/en15176276

[5] Groumpos PP. A critical historical and scientific overview of all industrial revolutions. IFAC-PapersOnLine. 2021;**54**(13):464-471. DOI: 10.1016/j.ifacol.2021.10.492

[6] Clim A, Toma A, Zota RD, Constantinescu R. The need for cybersecurity in industrial revolution and smart cities. Sensors. 2023;**23**(1). DOI: 10.3390/s23010120

[7] Pizoń J, Gola A. Human–machine relationship—Perspective and future roadmap for industry 5.0 solutions. Machines. 2023;**11**(2). DOI: 10.3390/machines11020203

[8] Papulová Z, Gažová A, Šufliarský L. Implementation of automation Technologies of Industry 4.0 in automotive manufacturing companies. Procedia Computer Science. 2022;**200**(2019):1488-1497. DOI: 10.1016/j.procs.2022.01.350

[9] Jämsä-Jounela SL. Future trends in process automation. Annual Reviews in Control. 2007;**31**(2):211-220. DOI: 10.1016/j.arcontrol.2007.08.003

[10] Pacaux-Lemoine MP, Trentesaux D, Zambrano Rey G, Millot P. Designing intelligent manufacturing systems through human-machine cooperation principles: A human-centered approach. Computers and Industrial Engineering. 2017;**111**:581-595. DOI: 10.1016/j.cie.2017.05.014

[11] Sharp M, Ak R, Hedberg T. A survey of the advancing use and development of machine learning in smart manufacturing. Journal of Manufacturing Systems. 2018;**48**:170-179. DOI: 10.1016/j.jmsy.2018.02.004

[12] Leng J et al. Industry 5.0: Prospect and retrospect. Journal of Manufacturing Systems. 2022;**65**(September):279-295

[13] Javaid M, Haleem A, Singh RP, Rab S, Suman R. Significant applications of Cobots in the field of manufacturing. Cognitive Robotics. 2022;**2**(October):222-233. DOI: 10.1016/j.cogr.2022.10.001

[14] Haleem A, Javaid M, Singh RP, Rab S, Suman R. Hyperautomation for the enhancement of automation in industries. Sensors International. 2021;**2**(August). DOI: 10.1016/j.sintl.2021.100124

[15] Fogliatto FS, Da Silveira GJC, Borenstein D. The mass customization

decade: An updated review of the literature. International Journal of Production Economics. 2012;**138**(1):14-25. DOI: 10.1016/j.ijpe.2012.03.002

[16] Paturi UMR, Cheruku S. Application and performance of machine learning techniques in manufacturing sector from the past two decades: A review. Materials Today: Proceedings. 2020;**38**:2392-2401. DOI: 10.1016/j.matpr.2020.07.209

[17] Dencker K, Fasth Å, Stahre J, Mårtensson L, Lundholm T, Akillioglu H. Proactive assembly systems-realising the potential of human collaboration with automation. Annual Reviews in Control. 2009;**33**(2):230-237. DOI: 10.1016/j.arcontrol.2009.05.004

[18] Xu X, Lu Y, Vogel-Heuser B, Wang L. Industry 4.0 and industry 5.0—Inception, conception and perception. Journal of Manufacturing Systems. 2021;**61**(September):530-535. DOI: 10.1016/j.jmsy.2021.10.006

[19] Tercan H, Meisen T. Machine learning and deep learning based predictive quality in manufacturing: A systematic review. Journal of Intelligent Manufacturing. 2022;**33**(7):1879-1905. DOI: 10.1007/s10845-022-01963-8

[20] Lee J, Davari H, Singh J, Pandhare V. Industrial artificial intelligence for industry 4.0-based manufacturing systems. Manufacturing Letters. 2018;**18**:20-23. DOI: 10.1016/j.mfglet.2018.09.002

[21] Cioffi R, Travaglioni M, Piscitelli G, Petrillo A, De Felice F. Artificial intelligence and machine learning applications in smart production: Progress, trends, and directions. Sustainability. 2020;**12**(2). DOI: 10.3390/su12020492

[22] Liu Q, Chen C, Chen S. Key technology of intelligentized welding manufacturing and systems based on the internet of things and multi-agent. Journal of Manufacturing and Materials Processing. 2022;**6**(6). DOI: 10.3390/jmmp6060135

[23] Abdul Kadir MH, Asmelash M, Azhari A. Investigation on welding distortion in stainless steel sheet using gas tungsten arc welding process. Materials Today: Proceedings. 2020;**46**:1674-1679. DOI: 10.1016/j.matpr.2020.07.264

[24] Geng Y, Lai M, Tian X, Xu X, Jiang Y, Zhang Y. A novel seam extraction and path planning method for robotic welding of medium-thickness plate structural parts based on 3D vision. Robotics and Computer-Integrated Manufacturing. 2023;**79**(May 2022). DOI: 10.1016/j.rcim.2022.102433

[25] Tsuzuki R. Development of automation and artificial intelligence technology for welding and inspection process in aircraft industry. Weld World Enterprise. 2022;**66**(1):105-116. DOI: 10.1007/s40194-021-01210-3

[26] Rout A, Deepak BBVL, Biswal BB. Advances in weld seam tracking techniques for robotic welding: A review. Robotics and Computer-Integrated Manufacturing. 2019;**56**(September 2018):12-37. DOI: 10.1016/j.rcim.2018.08.003

[27] He K, Xia Z, Si Y, Liang J, Yong J, Shi W. Detection of arc characteristics and weld forming quality of aluminum alloy DP-MIG welding using AE signal through resonance demodulation. Journal of the International Measurement Confederation. 2022;**189**(November 2021). DOI: 10.1016/j.measurement.2021.110427

[28] Shen W, Hu T, Zhang C, Ye Y, Li Z. A welding task data model for intelligent

process planning of robotic welding. Robotics and Computer-Integrated Manufacturing. 2020;**64**(September 2019). DOI: 10.1016/j.rcim.2020.101934

[29] Chuang TC, Lo YL, Tran HC, Tsai YA, Chen CY, Chiu CP. Optimization of butt-joint laser welding parameters for elimination of angular distortion using high-fidelity simulations and machine learning. Optics and Laser Technology. 2023;**167**. DOI: 10.1016/j.optlastec.2023.109566

[30] Kesse MA, Buah E, Handroos H, Ayetor GK. Development of an artificial intelligence powered tig welding algorithm for the prediction of bead geometry for tig welding processes using hybrid deep learning. Metals (Basel). 2020;**10**(4). DOI: 10.3390/met10040451

[31] Wang B, Hu SJ, Sun L, Freiheit T. Intelligent welding system technologies: State-of-the-art review and perspectives. Journal of Manufacturing Systems. 2020;**56**(July):373-391. DOI: 10.1016/j.jmsy.2020.06.020

[32] Carvalho N, Chaim O, Cazarini E, Gerolamo M. Real-time monitoring system to lean manufacturing. Procedia Manufacturing. 2018;**22**:533-538. DOI: 10.1016/j.promfg.2018.03.078

Chapter 2

State-Space Modeling of Weld Bead Geometry in the Gas Metal Arc-Direct Energy Deposition Process Applied to Wire and Arc Additive Manufacturing and Welding Processes

Jairo José Muñoz Chávez,
Margareth Nascimento de Souza Lira and
Sadek Crisostomo Absi Alfaro

Abstract

One of the main problems of additive manufacturing with electric arc and welding, in general, is the difficulty in controlling or predicting the output variables and their parameters, as well as creating a model that effectively represents the changes in the main variables involved in the system. These changes during the deposition process can promote the formation of splashes, instabilities, and changes in the geometry of the beads, making the analysis of these variables important, as it will be through them that the quality of the deposit and the desired characteristics will be established. Despite the correlation between the variables, they present nonlinear and chaotic behavior. With this, the purpose of this research is mathematical modeling in state space that allows an approximation to the model in state spaces, an approximation of the real values of the process, and a knowledge of the system composed of a set of input, output, and states related to each other by means of first-order differential equations. The model was validated from depositions *via* a design of experiments with central composite planning monitored with the use of sensors to capture the characteristics of the beads (e.g., molten pool, width, penetration, and height).

Keywords: additive manufacturing, welding, GMA, prediction, mathematical modeling

1. Introduction

The model constitutes an abstract representation of a certain aspect of reality. In its structure, the elements that characterize the modeled reality intervene in the existing relationships between them. A mathematical model is a model based on mathematical logic, whose elements are essentially variables and functions, and the relationships between them are expressed through equations, inequalities, logical operators, state space, vectors, or matrices that correspond to the real relationships that can be modeled (technological relationships, physical laws, forces, time-varying population, financial, or market quantities) [1].

Some of the models presented in the literature are complex, having an exacerbated number of input data and difficult-to-solve equations, thus promoting, for control or identification, high processing time, and high computational cost [2]. Other simplified models present shorter processing times, but do not approach the reality of the phenomenon involved as required. This becomes a problem when you want to obtain acceptable information with a high speed in data processing. The main objective in predicting any physical phenomenon and an important part of the present work is to develop a mathematical model that allows describing and predicting the behavior of this phenomenon and preferably with simplified equations [3]. In the present work, it is important to learn and mathematically model the phenomena that involve the physics of the arc in the welding process, such as, the values of reinforcement width and penetration using the GMAW welding technique [4, 5].

For control engineering, any dynamic, time-varying, or time-invariant physical system can have a state-space representation. The state-space model is a mathematical model of a physical system composed of a set of input, output, and state variables related to each other through first-order differential equations. Variables are expressed in vectors and differential and algebraic equations can be written in matrix form when the dynamical system is linear and time invariant. The state-space representation with a time-domain approach provides a practical and compact way to model and analyze systems with multiple inputs and outputs. In the frequency domain, the use of the state-space representation is not limited to systems with linear components and zero initial conditions. The "state space" refers to the space whose axes are the state variables. The state of the system can be represented as a vector within this space [1].

2. State-space modeling

The proposal in this item is to develop a mathematical model by state space, that is, a dynamic model that can: predict the desired output variables; generalize to different operating points in working ranges in globular and droplet mode; and predict the proposed outputs, width, reinforcement and penetration through input values by the state variables used for control.

2.1 Definition of variables

In order to achieve a complete study of the phenomenon, eight states proposed for output prediction were used: width, reinforcement, and penetration. Next, **Table 1** describes the proposed states, the type of measurement made in each of the variables

Symbol	Variable	Unit	Measurement type	Sensor or technique used
X1: *ls*	Stick out	(mm)	Offline	Shadowgraph
X2: Ia	Current	(A)	Online	Ammeter
X3: Ws	Wire feed speed	(m/min)	Online	Encoder on feeder motor
X4: Cs	Solid wire length	(mm)	Online	Shadowgraph
X5: Ts	Travel speed	(mm/s)	Online	Encoder in linear table motor
X6: Wb	Bead width	(mm)	Online	• Infrared Camera • Scanner
X7: R	Reinforcement	(mm)	Online	• Infrared Camera • Scanner
X8: P	Penetration	(mm)	Online	Macrograph

Table 1.
State variables with the type of measurement and the sensor with which it was measured.

Variable	U1—Uoc	U2—Uma	U3—Ums
Detailing	Voltage applied by the source (V)	Voltage applied across motor armature to obtain Ws (V)	Voltage applied to motor armature to obtain Ts (V)
Measurement	On-line	On-line	On-line
Sensor	Voltmeter	Voltmeter—Ratio between the load on the motor and the wire feed speed	Voltmeter—Relationship between the load on the motor and the speed of the linear displacement table that contains the part to be welded

Table 2.
Input variables.

(Online or Offline), the units used, and the sensor or sensors with which the measurement was performed.

The input variables for the control are U1, U2, and U3 and their details are shown in **Table 2**.

For the purposes of formulating the mathematical model by state space, a first study is presented, based on equations taken from other works in the field of arc physics and some considerations of our own. For this case, using a mathematical model, there is greater complexity when working in short circuit together with other transfer modes, which is why in this research proposal the droplet mode will be the focus, given the difficulty of working with several modes of transfer in the same model, considering models developed for GMAW such as the Moore, Naido and Tayler model [6], Thomsen [7] and Plankaert, Djermoune, Brie and Richard [8]. The plant of the system can be proposed for this research as shown in **Figure 1** and the variables detailed in **Table 3**.

The current presents different oscillations, the explanation of these oscillations working with the source at constant voltage is due to the growth of the drop and changes in the conduction zone. This is because for higher values of current and fusion energy, the drop size decreases and its detachment frequency increases. Various techniques were used to see changes in arc length, droplet size, and droplet

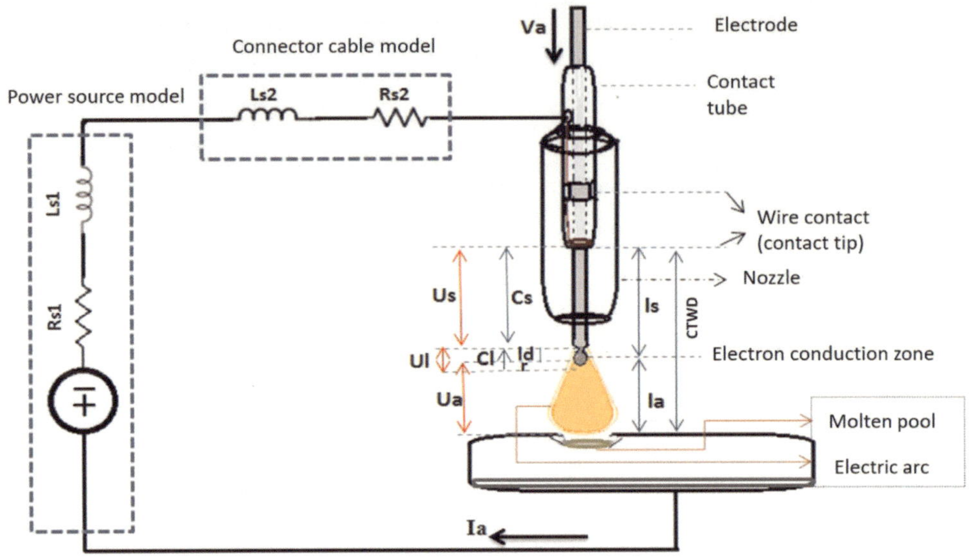

Figure 1.
Welding plant and its main electrical components.

detachment frequency. Many of these techniques use infrared cameras, optical filters, digital filters, high-speed cameras, algorithms for data processing, among other tools. Making use of experimental data collected in the tests, based on images and videos taken with the different techniques mentioned, an analysis was made of how these variables change, such as the length of arc and wire, subsequently managing to relate them to the input and output parameters when proposing the state equations; therefore, in this first step, the influence of the growth and detachment of the drops in the total system will be introduced and some equations are proposed, which will help to understand the proposed state system. For this, was approached an analysis of droplet formation and arc length.

2.2 Analysis of droplet formation and arc length

In the first observations when performing changes in voltage, current, and wire speed, the videos show records of a change in droplet size and frequency of droplet detachment. The reverse also happens where changing the droplet and its frequency affects the current, due to changes in the arc length and the resistance in the arc that the droplet generates. Thus, for higher values of current and fusion energy, the drop size decreases and its detachment frequency increases. This causes the current signal to also present a smaller oscillation in its signal intensity changes, but happening with greater frequency.

In most models, values work where the radius of the drop is proportional to the radius of the wire, that is, in the transition from globular to droplet, where the conduction zone (h) is below the center of the drop, at a distance $h = re.\cos\theta$, where re is the electrode radius. But for higher values of current and voltage above the transition point, the droplet has a smaller size, and at the same time increases the anodic zone in the electrode and the conduction zone, with which this conduction zone appears above the center of the drop as evidenced in some images of the welding

Variable	Description
l_a	Arc length
CTWD	Contact to work distance
l_s	Stick out
Cs	Wire length in solid state
Cl	Wire length in liquid state
ld	Drop Offset—distance from the solid wire to the center of the drop
h	Distance from the center of the drop to the conduction zone on the Y axis = $r_d \cos(\theta)$
r	Distance from droplet tip to conduction zone
re	Electrode radius
rd	Drop radius
θ	W.t; where t is time and W is a constant representing the angular frequency in rad/s which relates to various factors that determine drop growth.
Us	Voltage drop in solid wire
Ua	Arc voltage drop
Ul	Voltage drop in liquid wire
Uoc	Open circuit voltage drop = sum of voltage drops in the system
Ls1	Inductance generated at the source
Ls2	Inductance generated in the cable
Rs1	Resistance inside the source
Rs2	Cable resistance
Ia	Current flowing in the cable
Ws	Wire feed speed
d	Length between the end of the wire and the part

Table 3.
Welding plant variables used to formulate the mathematical model.

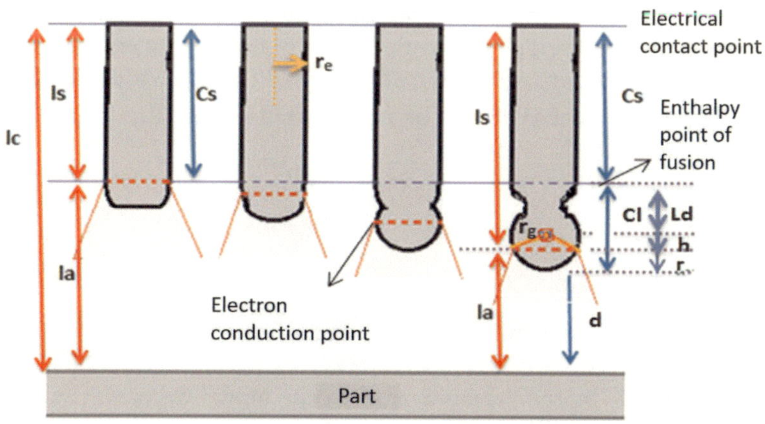

Figure 2.
Behavior of the conduction zone during droplet formation at the globular transition point—Droplet with droplet equal to the wire size.

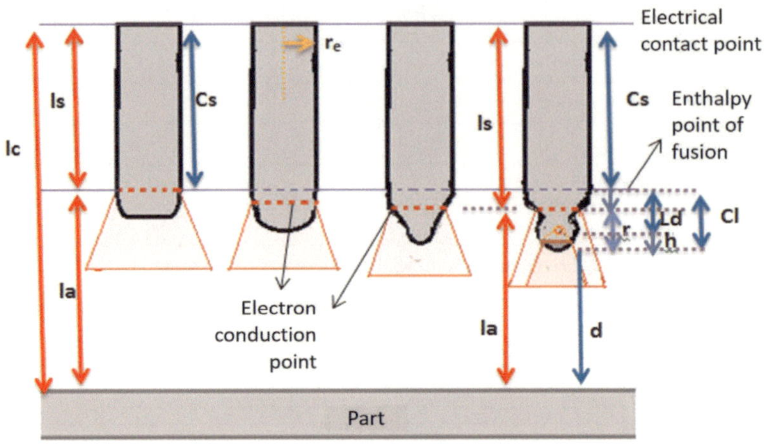

Figure 3.
Behavior of the conduction zone during drop formation for high current and voltage values.

process shown in **Figures 2** and **3** and visualized in the images captured experimentally in **Figure 4** by different capture methods. Trying to determine the geometry of the plasma, the length of the arc and the location of the wire and drop. Another point to consider is the high frequency in the formation of drops, this short time makes the oscillation in the current to be minimal, as well as its arc length.

Changes in arc length due to droplet formation introduce noise into the system due to the change in arc voltage drop. The relationships between the lengths are (Eqs. (1)–(3)):

$$l_a = CTWD - ls \tag{1}$$

$$ls = Cs + l_d + \mathrm{h} = Cs + l_d + r_d \cos(\theta) \tag{2}$$

$$l_a = d + r \tag{3}$$

This value of r can be considered approximately equal to the displacement of the droplet in droplet mode. The electrode length in the solid state shows dynamic variation dependent on the wire feed speed (Ws), in addition to the melt volume rate ($\dot{VfR}$). Such behavior can be represented below (Eq. (4)).

$$\frac{dCs}{dt} = W_s - \frac{\dot{VfR}}{\pi r_e{}^2} \tag{4}$$

Where:

$\dot{VfR}$ = Melted volume per unit time.

Ws = Wire feed speed.

Substituting Eq. (4) by the variables of the state model we have (Eq. (5):

$$\frac{dX4}{dt} = \mathrm{X3} - \frac{\dot{VfR}}{\pi r_e{}^2} \tag{5}$$

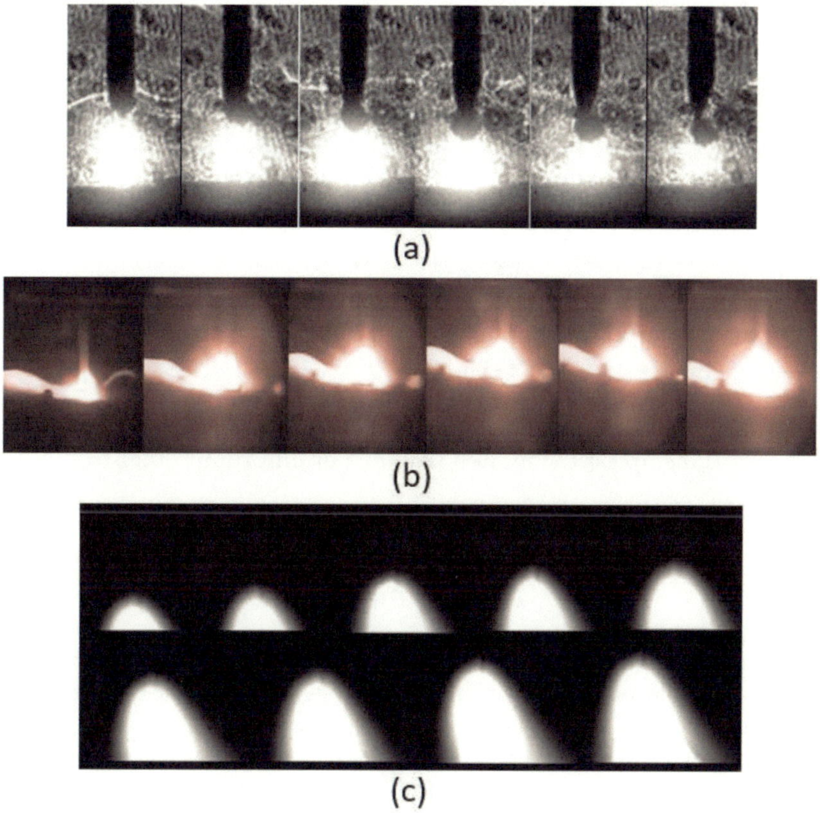

Figure 4.
(a) Plasma and wire visualized with shadowgraph at high voltages during drop formation and detachment. (b) Voltage increments for viewing the plasma with the infrared camera. (c) Voltage increments to visualize plasma only using optical bandpass filter between 625 and 700 nm and a high-speed camera.

The electrode is molten, constantly adding liquid metal to the drop that is forming at the tip of the electrode. The change in the mass of the drop (m_d) is directly influenced by the melting rate of the electrode, as evidenced in the following equation (Eq. (6)):

$$\frac{dm_d}{dt} = \dot{V}fR * \rho_e \tag{6}$$

The length of the electrode (h) between the torch and the weld pool depends on the difference between the wire feed speed and the molten portion of the wire when a certain voltage is applied, which allows to maintain a balance between the molten material rate and the length distance of bow (Eq. (7)).

$$\frac{dh}{dt} = \frac{\dot{V}f_R}{A} \text{ ou } h = \frac{Vf_R}{A} \tag{7}$$

Where $\dot{V}fR$ is the volumetric melting rate of the electrode per unit time and A constitutes the area of the electrode (Eq. (8)).

$$A = \pi {r_e}^2 \tag{8}$$

Therefore, when dividing $\dot{Vf}R$ using area A, a value for the length of molten wire per unit of time (h) is obtained. The melting rate of the electrode is due to the influence of two factors, namely: ohmic and anodic heating. The melting rate of the electrode varies depending on the intensity of the current, the type of material, and the gas used, and this results in an equation that depends on the current and that for each material and gas used presents constants C1 and C2 (Eqs. (9) and (10)).

$$Vf_R = C_1 I + C_2 I^2 \rho l_s \tag{9}$$

Substituting for the state variables, the result is:

$$Vf_R = C_1 X_2 + C_2 {X_2}^2 \rho X_1 \tag{10}$$

Another relationship for the molten volume per unit time comes from Eq. (11) or Eq. (12):

$$\mathrm{d}\, Vf_R = A.W_s.dt \tag{11}$$

$$Vf_R = A.W_s.t \tag{12}$$

Where A is the area of the electrode or wire. The volume melted per unit time always equals the volume added per unit time. When you increase or decrease the voltage and current, you increase or decrease the arc length, compensating for the heat to fuse the wire and stabilizing the entire system. There is a great influence of the inductance that is related to the variation of the current and the applied voltage, as well as a relationship of the resistance of the given source and the current that is circulating in the system as exposed in the Eq. (13).

$$L_s \frac{dI}{dt} + (R_f + R_a)I + U_a = U_{oc} \tag{13}$$

Where:

Ls = Sum of total inductance in the system $L_s = \frac{Ls1 * Ls2}{Ls1 + Ls2}$.

Rf = System resistance without taking into account the arc resistance.

Rf = R_{s1} + R_{s2}.

Rs = Rf + Ra.

Ua = Arc voltage drop.

Uoc = Voltage at the source output.

Arc voltage and resistance are related to voltage drops. A strong relationship between arc voltage, arc length, and arc current is also shown. In the transition range from globular to droplet mode, where the drop has a radius equal to the size of the wire radius, the conduction zone was calculated performing an approximation only with the volumes and calculating the value of the conduction wire length (lac). For this relationship, the volume of the wire is considered as a cylinder, the spherical drop with a radius equal to the radius of the cylinder with a ratio between the volume of the drop and the volume of the cylinder of 3/2 (**Figure** 5).

Based on **Figure** 5 and previous analyses, Eq. (14):

$$\frac{1}{3}.V_g = A * l_{ac} \tag{14}$$

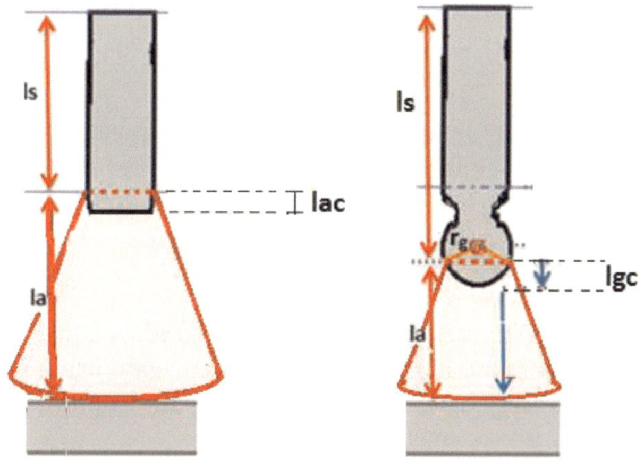

Figure 5.
Scheme of the plasma shape during the formation of a droplet at the transition point (when rg = re droplet mode), with arc length variation (la) and conduction zone.

Respecting V_g is the volume of the drop and that the voltage and current have values where the radius of the drop is equal to the radius of the electrode and later isolating l_{ac} we get the value of the wire length of the conduction zone at the transition point, $l_{ac} = 0.267\ mm$. Experimentally, it is observed that there is no linear or quadratic variation in l_{ac} as the voltage increases, so it is proposed to apply a 3/2 exponent that matches the experimental data and add the corresponding constants of proportionality. With the application of these proposed equivalences, the equation for the length of the wire in electrical conduction finaly looks like this (Eq. (15)):

$$l_{ac} = \frac{\left(C7 * e^{0.38 * (Uoc - Ut)^{\frac{1}{3}}} - C9 * l_a^{\frac{3}{2}} * r_e - C10\right)}{\pi * {r_e}^2} \tag{15}$$

Where Uoc is the output voltage of the source, Ut is the voltage at the transition point for this wire and gas used, which experimentally obtained Ut - 26.5 V for pure argon, and for argon with 4% CO2 Ut = 23 V. Then continue using the values with 4% CO2. And finally the constants C7, C9, and C10, are system constants that have been found experimentally. Substituting the constants and the Ut value, the conduction length finally looks like this (Eq. (16)):

$$l_{ac} = \frac{\left(e^{0.38 * (Uoc - 23)^{\frac{1}{3}}} - 0{,}032 * l_a^{\frac{3}{2}} - 0.42\right)}{\pi * {r_e}^2} \tag{16}$$

It should be noted that these equations are only valid for the globular and droplet modes. Now, we analyze the dynamics of the droplet detachment frequency and its volume by the melting mass. Equating the volume of the drop with the volume of the molten liquid wire when rg = re (Eq. (17)):

$$Vg = \frac{4}{3}\pi * {r_e}^3 = \pi * {r_e}^2 * l \tag{17}$$

Where the length of cast wire correspondig to the drop is (Eq. (18)):

$$l = W_s * \Delta t \tag{18}$$

Isolating Δt and replacing values of l *and* W_s at the transition point, where the radius of the drop equals the radius of the electrode (rg = re) one arrives at (Eq. (19)):

$$\Delta t = \frac{4 * 0.6\,mm}{3 * 113.34\,mm/seg} = 7.058\,ms \tag{19}$$

Therefore, it takes 7.058 ms to form a drop with a radius equal to the wire using a wire speed equal to 6.8 m/min. Performing regression again using experimental data, a relationship was found for changes in the frequency of drop formation with the variation in voltage and it was arrived at (Eq. (20)):

$$\Delta t_f = t = \frac{10.5\,ms}{(Uoc - 21)^{7/12}} \tag{20}$$

Isolating re from Eq. (17) and replacing l by Eq. (18), also replacing re with rg and $\Delta t = \Delta t_f$ the drop radius equation is obtained for different wire speeds (Eq. (21)).

$$r_g = \frac{3}{4} * \text{Ws} * \Delta t_f \tag{21}$$

Where: Δt_f is the maximum drop formation time.

Figure 6 below shows the growth period and maximum droplet radius size along with the detachment frequency for each stress value.

The drop is ranging from (½ re) to (1 re +2 rg). With these relations, it is possible to describe the solid wire and cast wire equation and correlate them with the following equations (Eqs. (22) and (23)):

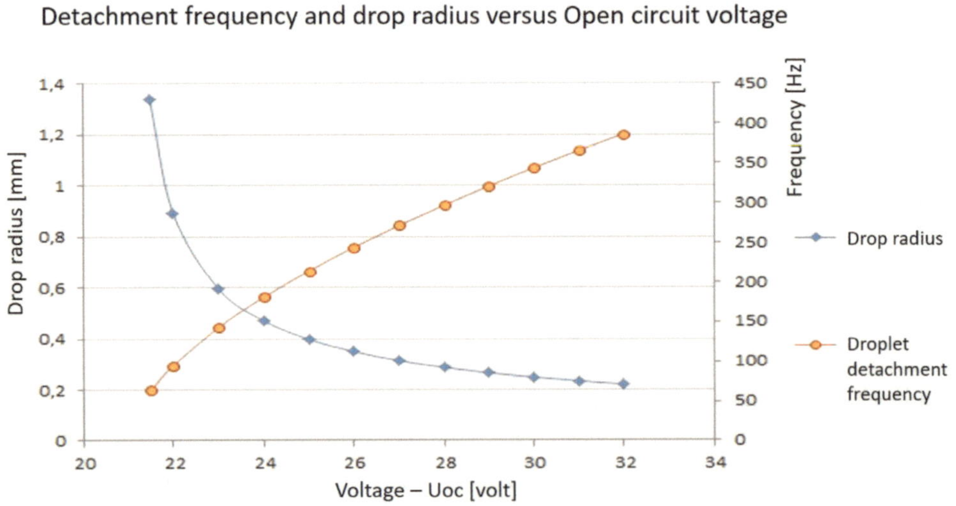

Figure 6.
Droplet detachment frequency and size with open-circuit voltage variations.

$$\text{la} = 15\text{–ls–lac} \tag{22}$$

$$\text{Cs} = \text{ls} + \text{lac–Cl} \tag{23}$$

Where Cl is the length of cast wire.

To model the detachment of the drop, the application of the Fourier series was proposed for a saw-type wave, and using the first 4 terms of sine, together with the equations described above for the period and volume of the drop, Eq. (24) below to model droplet size variation and cast wire length (Cl). Where:

$$Cl = \left(\tfrac{1}{2}r_e + 2r_g\right) * \left[\frac{1}{\pi} * \left[sen\left(x - \frac{\pi}{2}\right) - \frac{1}{2}sen\left(2x - \frac{\pi}{2}\right) + \frac{1}{3}sen\left(3x - \frac{\pi}{2}\right) - \frac{1}{4}sen\left(4x - \frac{\pi}{2}\right)\right] + 1 + \frac{1}{2}r_e\right] \tag{24}$$

Because: X = W*t, and X is cyclic, then X = 2Π/Δt. Substituting r_e, r_g, X and Δt in the equation we get (Eq. (25)):

$$Cl = \left(0.3 + \frac{3}{2} * Ws * \frac{10,5}{(Uoc - 21)^{7/12}}\right) * \frac{1}{\pi} * \left[sen\left(\frac{2\pi(Uoc - 21)^{7/12}}{10,5} - \frac{\pi}{2}\right) - \frac{1}{2}sen\left(\frac{4\pi(Uoc - 21)^{7/12}}{10.5} - \frac{\pi}{2}\right) + \frac{1}{3}sen\left(\frac{6\pi(Uoc - 21)^{7/12}}{10.5} - \frac{\pi}{2}\right) - \frac{1}{4}sen\left(\frac{8\pi(Uoc - 21)^{7/12}}{10.5} - \frac{\pi}{2}\right) + 1 + \frac{1}{2} * r_e\right] \tag{25}$$

To facilitate the handling of this equation (Eq. (25)) in the state model, it will be placed from now on as:

$$Cl = \left(0.3 + \frac{3}{2} * Ws * \frac{10,5}{(Uoc - 21)^{7/12}}\right) * \frac{1}{\pi} * [***] \tag{26}$$

Where the expression [***] is the Fourier expression belonging to the sines.

Using training values with input variables such as source output voltage (Uoc) and wire feed speed Ws, one obtains the drop size value and detachment frequency shown in **Figure 7** below.

Performing an analysis of the shape of the plasma in the captured images, where a source of direct current with reverse polarity was used, it presented a bell shape, as expected according to the references. The energy released in the displacement of electrons is given by E = I*V*Δt, the voltage drops define areas where there is more consumption of this energy. Part is dissipated in cables and connections, but most is being consumed in cathode, anode, and plasma. This energy can be visualized by the isotherms, in **Figure 8**.

The isotherms show temperature gradients from the plate to the tip of the wire in the Y axis and from the outside to the center of the arc in the XZ direction radially around a circle. The T° gradient in the XZ and Y isotherms close to the part has a relationship with the gradient in penetration. We first analyze the relationship

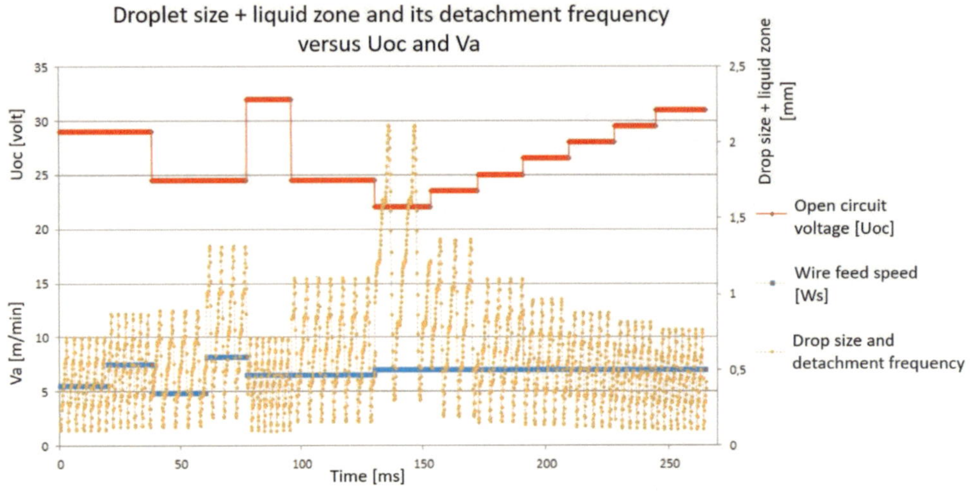

Figure 7.
Different Cl sizes and detachment frequencies simulated for open-circuit voltage variations.

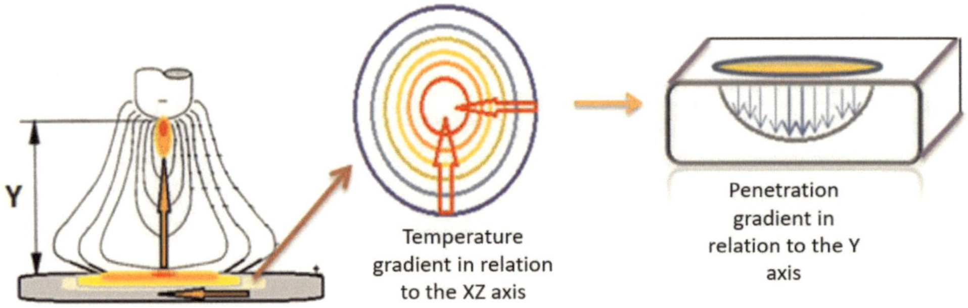

Figure 8.
Scheme of isotherms and temperature gradients and the relationship with the penetration gradient.

between the width and the energy contained in the plasma volume to propose an equation for the state-space model. There is a total energy of the system and an energy consumed in the plasma expressed in the following equations (Eqs. (27) and (28)):

$$Et = I * V * dt \tag{27}$$

$$Ep = C8 * V_{pl} * dt \tag{28}$$

Where Vpl is the plasma volume approximated to the volume of a cone and with some other considerations one arrives that the plasma volume equation is (Eq. (29)):

$$Vpl = \frac{1}{12} * \pi * La^2 * (l_a + r_e) \tag{29}$$

Therefore (Eq. (30)):

$$I * V = C8 * \frac{1}{12} * \pi * La^2 * (l_a + r_e) \tag{30}$$

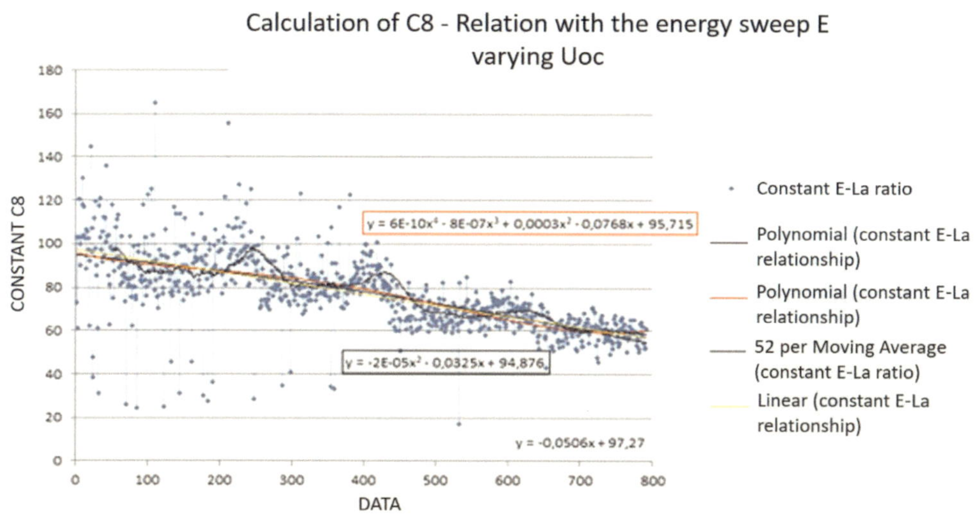

Figure 9.
Calculation of equations that model the variation of C8 with the variation of power or consumed energy.

Where La is the bead width and la is the arc length. Isolating C8 and replacing the values of I, V, La, and l_a by experimental values in droplet mode, we obtain the value of C8. The result shows a trend, but due to the dispersion of the data, a linear and polynomial approximation of degree 2 is performed. The tendency of C8 to decrease with the proportional increase of Uoc every 200 data points is shown in **Figure 9**. It is also observed how the dispersion of the values is smaller with higher voltages, indicating greater stability of the system when increasing the voltage.

Using an additional linear relationship in the equation in the work range for dropleting and with experimental values, the constant C8 = 90 was found and isolating La we obtain (Eqs. (31) and (32)):

$$La^2 = \frac{12 * I * Uoc}{\pi * (l_a + r_e) * (90 - (Uoc - 26,5) * 10)} \tag{31}$$

$$La = \sqrt{\frac{12 * I * Uoc}{\pi * (l_a + r_e) * (90 - (Uoc - 26,5) * 10)}} \tag{32}$$

Where to CTWD = 15 if have: l_a = 15 – l_s and l_s = X1.

Was used experimental data to test the proposed equation (Eq. (32)). Where current and voltage in the arc are noisy because of droplet formation and detachment as well as changes in conduction zone and arc length mentioned earlier. The proposed equation, even with this noise, shows a tendency in the prediction of the width as shown in **Figure 10**. But using all the equations of state with adjustments and without using the experimental current values, the prediction improves a lot.

Substituting the notation of the variables to use in the state model, we get (Eqs. (33) and (34)):

$$X6 = \left(\frac{12 * U1}{\pi * (90 - (U1 - 26,5) * 10)}\right)^{½} \left(\frac{X2}{(15 - X1 + r_e)}\right)^{½} \tag{33}$$

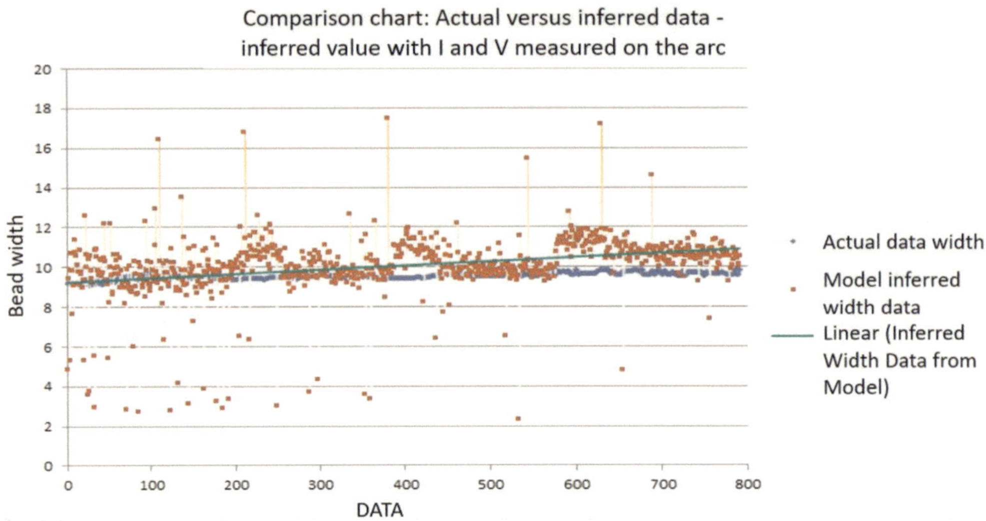

Figure 10.
Previous test of the width prediction values of the X6 equation of state, using experimental data of voltage and current in the arc at the transition point and comparing with the actual width data.

$$\dot{X6} = \left(\frac{12 * U1}{\pi * (90 - (Uoc - 26,5) * 10)}\right)^{½} * \frac{1}{2}\left(\frac{X2}{(15 - X1 + r_e)}\right)^{-½} * \frac{\dot{X2} * (15 - X1 + r_e) - \dot{X1} * X2}{(15 - X1 + r_e)^2} \quad (34)$$

The other equation of state to determine the value of reinforcement and the relation with the width comes from the relation of volumes, the molten volume of the wire (Vfr) and the deposited volume (Vd) which depends on the welding speed and the cross section of the cord per unit time. These volume expressions are expressed below (Eqs. (35) and (36)):

$$Vf = A1 * Ws * dt = A1 * Ws * \Delta t \quad (35)$$

$$Vd = Vf = A2 * Ts * dt = A2 * Ts * \Delta t \quad (36)$$

Equating the equations (Eq. (37)):

$$A1 * Ws = A2 * Ts \quad (37)$$

Where:
$A1 = \pi * {r_e}^2$;
$A2 = \pi * \frac{a * b}{2}$;
$a = R$;
$b = La/2$;
Replacing and isolating the reinforcement R (Eq. (38)):

$$R = \frac{4 * {r_e}^2 * Ws}{La * Ts} \quad (38)$$

Substituting the reinforcement equation for the state variables, we have (Eq. (39)):

$$X7 = \frac{4 * {r_e}^2 * X3}{X6 * X5} \tag{39}$$

The last proposed equation of state is the relationship between penetration and other variables, which will depend on constants specific to the material, such as material density, thermal conduction, specific heat, enthalpy of fusion, electrical conductivity, electrical resistance, thermal energy, arc voltage drop, anodic voltage drop, thermal dissipation by conduction, convection and radiation, exposure time of the arc with the puddle which will depend on the welding speed, and area of contact of the plasma with the piece which will depend on the length of arc and this in turn of the wire speed and the tension. Due to the large number of variables, it was decided to simplify the equation and leave the variables with the greatest influence on penetration. After analyzing the energy required to melt the metal pool, and the energy in the plasma created by the current, the following equation is proposed (Eq. (40)):

$$P = \left(\frac{Cf * C5 * I * Ua}{\rho * \pi * \left(\frac{La}{2}\right)^2 * Ts} \right) \tag{40}$$

Leaving only one constant $C9 = \left(\frac{4 * Cf * C5 * C6}{\rho * \pi}\right)$; replacing is (Eq. (41)):

$$P = C9 * \left(\frac{I * Uoc}{La^2 * Ts} \right) \tag{41}$$

Substituting for the state variables, we get (Eq. (42)):

$$X8 = C9 * \left(\frac{X2 * U1}{X6^2 * X5} \right) \tag{42}$$

Finally, replacing lac and Cl in Eq. (22) we get Eq. (42) for Cs (Eq. (43)).

$$Cs = ls + \frac{\left(e^{0.38 * (Uoc-23)^{\frac{1}{3}}} - 0.032 * l_a^{\frac{3}{2}} - 0.42\right)}{\pi * {r_e}^2} - \left(0.3 + \frac{3}{2} * Ws * \frac{10,5}{(Uoc-21)^{7/12}}\right) * \frac{1}{\pi} * [* * *] \tag{43}$$

Substituting for the state variables and knowing that la = 15 –X1, we have (Eq. (44)):

$$Cs = X1 + \frac{\left(e^{0.38 * (Uoc-23)^{\frac{1}{3}}} - 0.032 * (15 - X1)^{\frac{3}{2}} - 0.42\right)}{\pi * {r_e}^2} - \left(0.3 + \frac{3}{2} * X3 * \frac{10,5}{(Uoc-21)^{7/12}}\right) * \frac{1}{\pi} * [* * *] \tag{44}$$

Below is **Table 4** with the constants used for the parameterization of the proposed model.

Other constants used in the state-space model such as C1 and C2 are multiplications of inductances, resistances, and constant coefficients that are repeated in the equations.

Nomenclature	Symbol	Value (unit)
Total resistance	Rs	$6.8 * 10^{-3}$ (Ω)
Total inductance	Ls	$306*10^{-6}$ (H)
Arc resistance	Ra	0.0237 (Ω)
Arc length voltage factor	Ea	400 ($V.m^{-1}$)
Electrode resistance	ɤe	0.43 ($\Omega.m^{-1}$)
Space permeability	Pe	$1.257 * 10^{-6}$ ($Kg.m.s^{-2}A^{-2}$)
Gravity	G	9.8 ($m. s^{-2}$)
Liquid electrode density	ρ_e	7800 ($Kg.m^{-3}$)
electrode radio	re	0.0012 (m)
Plasma Density	ρ_p	1.6 ($Kg.m^{-3}$)2)
Time constant of wire and table motors	τm	$50*10^{-3}$ (s)
engine profit	Km	1.0 ($mV^{-1} s^{-1}$)
Surface tension of the material	γ	1.3 ($N.m^{-1}$)

Table 4.
Constants used for the parameterization of the proposed model.

The model needs parameterization, for a more accurate prediction, but the model also contributes to a better understanding of the behavior of the system or plant. Next, write the projected equations for each state, with the simplifications (Eqs. (45)–(52)):

$$\dot{X1} = X3 - \left(\frac{C1}{\Pi re^2}X2 + \frac{C2}{\Pi re^2}X1.X2^2\right) \tag{45}$$

$$\dot{X2} = \frac{1}{\ell s}\left[U1 - \left\{Rs + r\left(\ell s + \frac{1}{2}.\left(\frac{3md}{4\Pi re}\right)^{\frac{1}{3}} + Xd\right) + Ra\right\}.X2 - Ea(lc - X1) - Uo\right](*) \tag{46}$$

$$\dot{X3} = \frac{1}{Tma}(KmaU2 - X3) \tag{47}$$

$$\dot{X4} = X1 + \frac{\left(e^{0.38*(U1-23)^{\frac{1}{3}}} - 0.032*(15 - \mathrm{X1})^{\frac{3}{2}} - 0.42\right)}{\pi * {r_e}^2} - \left(0.3 + \frac{3}{2}*X3*\frac{10.5}{(Uoc-21)^{7/12}}\right)*\frac{1}{\pi}*[***] \tag{48}$$

$$\dot{X5} = \frac{1}{Tms}(KmsU3 - X5) \tag{49}$$

$$\dot{X6} = \left(\frac{12*U1}{\pi*(90-(U1-26,5)*10)}\right)^{½}*\frac{1}{2}\left(\frac{X2}{(15-X1+r_e)}\right)^{-½}*\frac{\dot{X2}*(15-X1+r_e)-\dot{X1}*X2}{(15-X1+r_e)^2} \tag{50}$$

$$\dot{X7} = \frac{4*{r_e}^2\frac{1}{Tma}(KmaU2-X3)*X6*X4 - 4*{r_e}^2*\dot{X3}*\left(X6*\frac{1}{Tms}(KmsU3-X4)+(\text{"}\dot{}\text{"})*X4\right)}{X6^2*X4^2} \tag{51}$$

Where ("") is Eq. state 6 to $\dot{X6}$

$$\dot{X8} = C9 * \left(\frac{U1 * \dot{X2} * X6^4 * X5^2 - U1 * X2 * (X6^2 * \dot{X5} + 2X6 * X5)}{X6^4 * X5^2} \right) \tag{52}$$

3. Conclusion

This work had the proposal to develop a mathematical model by equation of states, a dynamic model that can generalize to the globular and droplet ranges and predict the proposed outputs, width, reinforcement and penetration, through input values for the state variables used to control. The use of real data (shadowgraph, infrared camera, high-speed camera) for model calibration allows guaranteeing the reliability of the predicted results when using the state equations.

Acknowledgements

To CAPES for the financial support and to the University of Brasilia for the necessary infrastructure to carry out the tests.

Conflict of interest

The authors declare no conflict of interest.

Author details

Jairo José Muñoz Chávez[1*], Margareth Nascimento de Souza Lira[1] and Sadek Crisostomo Absi Alfaro[2]

1 Universidade Federal do Rio de Janeiro, Rio de Janeiro, Brazil

2 Universidade de Brasília, Brasília, Brazil

*Address all correspondence to: jairojmch@metalmat.ufrj.br

References

[1] Meerschaert MM. Mathematical Modeling. 4th ed. San Diego: Elsevier; 2013. 364 p

[2] Farias FWC, Payão Filho JC, Oliveira VHP. Prediction of the interpass temperauture of a wire arc additive manufactured wall: FEM simulations and artificial neural network. Additive Manufacturing. 2021;**48**:1022387. DOI: 10.1016/j.addma.2021.102387

[3] Muñoz Chávez JJ. Prediction of Weld Bead Geometry in the GMAW Process Using a Model in the Space of States and Artificial Neural and Networks (SVM, Neurodifuse and Recurrent High Order Neural Networks). Brasilia: Distrito Federal. Universidade de Brasilia; 2020

[4] Scotti A, Ponomarev V. Soldagem MIG/MAG: melhor entendimento, melhor desempenho. 1nd ed. Sao Paulo: Artliber; 2008. 284 p. DOI: 8588098423

[5] Jonson JA, Carlson NM, Smartt H, Clark D. Process control of GMAW: Sensing of metal transfer mode. Welding Journal. 1991;**70**:1965-1972

[6] Moore K, Naidu D, Tyler J. Gas metal arc welding control: Part I. Materials Science Nonlinear Analysis theory Methods & Applications. 1997;**30**:3101-3111. DOI: 10.1016/S0362-546X(97)00372-6

[7] Thomsen JS. Advanced Control Methods for Optimization of Arc Welding. Aalborg: Aalborg Universitet; 2004

[8] Planckaert JP, Djermoune EH, Brie D, Briand F, Richard F. Modeling of MIG/MAG welding with experimental validation using an active contour algorithm applied on high speed movies. Applied Mathematical Modelling. 2010;**34**:1004-1020. DOI: 10.1016/j.amp.2009.07.011

Section 2

Modifications in the Manufacturing Process

Chapter 3

Material and Load Path Appropriate Joining Techniques for FRP/Metal Hybrid Structures

Holger Seidlitz, Lars Ulke-Winter, Lucas Ost and Felix Kuke

Abstract

Fiber-reinforced plastics (FRP) offer great lightweight construction potential. However, the anisotropic high-performance materials can only be fully utilized through the development of material-specific joining processes. A literature study shows that conventional methods such as screwing, riveting and bolting are unsuitable, since the load-bearing fibers are severed in the joining region. This leads to high-stress concentrations. To reduce these, a method is presented in which through holes are created in thermoplastic FRP by reorienting the fibers in this area around the point of disruption in accordance with the load path. For this purpose, the polymer matrix is softened locally by applying heat and penetrated with a needle or mandrel. Based on this, a technology for material-specific joining of FRP and metals has been developed in the form of thermomechanical flow drill joining. In this process, a mandrel forms a bush from the metal component and deflects the fibers of the locally softened organic sheet to suit the material. Cold metal transfer (CMT) pin welding is presented as another fully automatable joining process. In this method, the softened plastic component is penetrated with the welding wire, displacing the fibers in the joining area and realigning them to suit the load path.

Keywords: fiber-reinforced plastics, thermomechanical flow drill joining, cold metal transfer pin welding, load path, fiber orientation

1. Introduction

Lightweight design is an essential component in product development for reducing greenhouse gas emissions during component manufacture and especially during the use of moving components, thus making an important contribution to conserving resources and achieving climate targets. Fiber-reinforced plastics (FRP), especially carbon fiber-reinforced plastics (CFRP), provide great lightweight construction potential due to their high specific strengths and stiffnesses. For these reasons, the use of FRP is steadily increasing. The demand for carbon fibers (CF) has grown by nearly 60% from 58 kt in 2015 to 92 kt in 2021 [1].

In hybrid structural parts, concentrated loads are induced via the metal components. The CFRP components are used for load transfer. Thermoplastic FRPs can be integrated into existing process chains of metal constructions due to their reversible formability and thus have a significant advantage over thermoset systems. However, the lightweight potential can only be fully exploited by joining processes appropriate for the material.

The following strategies, for example, are pursued in the technological implementation of mixed designs with thermoplastic FRP and metals: Molding of the plastic onto the metal component in the molding process (IMA: In Mold Assembly) and direct bonding of the plastic to the metal component after the molding process (PMA: Post Molding Assembly); cf. e.g. [2, 3]. With the PMA technique, the metal and plastic components can be produced in separate manufacturing processes and subsequently joined by means of mechanical joining technology. The design of components in multi-material design has given significant impetus to joining technology, e.g. the combined riveting and adhesive bonding already used in car body construction, cf. e.g. [4]. In IMA processes, on the other hand, the joining elements are generally based on tight fit, e.g. by undercuts or holes in the metal component. Thus, the hybrid structure is produced simultaneously in the combined master molding and assembly process. The need for high-strength joints designed to withstand loads increases with the complexity of the components and the number of material groups used. In the design and manufacture of highly loaded lightweight structures in mixed construction with high-strength thin-walled metal and fiber composite components, PMA processes have major advantages over IMA techniques. This is mainly due to the significantly greater design freedom of PMA, which allows the structure to be designed according to the load path over a wide range [5].

Current joining methods, analyzed in a literature review, such as riveting and bolting require pre-hole cutting operations in which the load-bearing fibers are severed. Due to the high resulting notch stress concentrations, material thickening or heavy inserts are necessary, which is contrary to the lightweight concept. Following the literature study, this chapter describes thermomechanical flow drill joining (FDJ) and cold metal transfer (CMT) pin welding for joining thermoplastic FRP to metals. In these processes, the matrix is locally thermally softened in the joining zone and the fibers are displaced in the joining process and aligned to the load path. In this way, notch stresses are reduced and higher shear tensile strengths are achieved than with riveting, for example. In this way, even particularly high-performance continuous fiber-reinforced plastics can be joined appropriately for the material.

2. State of the art: joining technologies for FRP/metal hybrid structures

In the following, known methods for joining FRP and metals are summarized. For a detailed analysis of the different joining technologies see Ref. [5].

The integration of structural FRP components in the body shell by means of mechanical joining technology was successfully demonstrated e.g. in Ref. [6–10]. In addition to the various tests on basic components, the integration of large-area structural components, such as floor and bulkhead modules as well as high-performance ribs made of textile-reinforced plastics for space-frame design concepts, was one of the focal points of the investigations.

With regard to joining technology, methods that were originally developed for isotropic materials are often used for the FRP material group. A basic classification of these joining techniques can be made into detachable, partially detachable and non-detachable [4]. In this context, extensive studies and design information for mechanical connection systems [2, 7], adhesive connections [11, 12], welded connections [13–16] as well as snap-fit [4, 17] and contour connections [11, 12] are sufficiently described in the literature. In addition, a distinction can also be made according to the principle of action between the firmly bonded (e.g. adhesive bonding, welding), the force-fit (e.g. clamping, pressing) and the form-fit connections (e.g. pins, contour connection) [18]. A further subdivision is made with respect to the geometric design of the joining zone into point and linear or planar joining systems.

2.1 Bolt connections

Bolt connections, which also include pin and rivet connections, are of particular importance for mixed construction with fiber composites and metals. Notes on the design are described in detail in Refs. [11, 18], for example. Bolt connections can be designed as single- or double-shear connections, overlapping and shafted, and with a lug on one or both sides [18]. At higher loads, the asymmetrical arrangement of the joining parts in the single-shear variant induces an eccentricity moment at the joint, which loads the joining partners for bending/peeling and leads to canting of the fastener. The associated failure can be counteracted by a symmetrical and thus double-shear arrangement of the parts to be joined. However, the increased component weight caused by the material doubling at the joint is a disadvantage.

The high potential of fiber-reinforced plastic composites cannot be fully exploited with rivets and bolts. Unlike metals, where stress concentrations at the holes are largely reduced by plastic flow processes, the installation of the fasteners (machining drilling, cutting component of self-piercing rivets) at the load application area leads to the severing of load-bearing fibers [19]. This causes high-stress concentrations at the hole edges, which often cause delamination and inter-fiber failures. The design and dimensioning of riveted and bolted joints in FRP is therefore fundamentally different compared to mono-metal designs, see e.g. [20–22]. In order to exploit the high material potential of fiber-plastic composites, the fiber arrangement in the load application area must be designed to withstand the stresses. The advantageous loop principle allows the introduction of particularly high loads [7].

2.2 Screw connections

Screw connections are widely used in mixed construction methods in general mechanical engineering and in vehicle construction. Unlike bolt connections, the introduction of loads at the connection point is predominantly perpendicular to the laminate plane. Classification can be made with respect to thread, head, and shank forms. In mixed designs with metals and FRP, a distinction is made between metal-side and plastic-side joining. An internal thread is produced on the plastic side using, for example, the tapping process [23] or thread-forming screws [24]. Alternatively, metallic thread inserts can also be integrated in the fiber composite [17]. For metal-side joining, flow drilling joining systems have proved particularly successful [25]. Metal- and plastic-side direct screwing processes as well as thread inserts are characterized by simple fabrication and one-sided accessibility of the joining point [26, 27].

2.2.1 Plastic side direct screwing

Various manufacturers offer thread-forming metal screws that can be inserted into structurally integrated core holes. The widely used direct screw process is used for inexpensively manufacturable unfilled as well as short- or long-fiber-reinforced injection-molded components and has been investigated, for example, in Refs. [2, 28–30]. The pull-out strength of plastic-set direct screw connections is determined, among other things, by the shear strength of the thread, with plastic threads exhibiting comparatively low values. The high degree of flexibility in the design of plastic components makes it possible to mold load introduction points in the injection molding process that are suitable for the load and serve to increase the screw-in depth [31]. In this way, screw-in tubes can be used to increase the pull-out strengths for thread-forming metal screws and reduce relaxation-related preload losses [29]. In the case of textile-reinforced plastic components, such load introductions can only be formed to a limited extent. For thin-walled thermoplastic FRP components in the high-performance range, direct screwing with thread-forming metal screws is, therefore, unsuitable.

2.2.2 Threaded inserts

Inserts are used in plastic components when requirements are placed on the connection point with regard to easy maintenance and interchangeability. Compared to the plastic-side direct screw connection systems, the thread insert merely transfers introduced assembly and operating loads from the screw to the surrounding plastic component. Threaded inserts are, on the one hand, embedded in-situ during the production of the molded part and, on the other hand, subsequently pressed in by ultrasonic or heating of the insert, screwed in or integrated by expanding the insert when driving in a screw in the FRP component [2]. In this process, the threaded inserts are mounted in screw-in tubes or directly in the component [24]. Inserts fulfill an integration-based adapter function between the plastic component and the metallic screw. A uniform distribution of the screw forces can be achieved as a result of special inserts with functional surfaces adapted to the material for force-locking, form-locking, or material-locking anchoring in the plastic component, cf. e.g. [32–34]. Threaded connections with inserts make an economic contribution to the implementation of connection systems suitable for large-scale production and are therefore used in many mixed construction systems with FRP components. For complex and highly stressed mixed construction systems, threaded inserts are not recommended—especially for thin-walled FRP components—because the connection forces are induced predominantly perpendicular to the direction of reinforcement. The transverse forces can thus only be applied locally with the connection system and thus cause interlaminar shear stress peaks at the joint, which often lead to premature failure of the FRP component.

2.2.3 Metal-side direct screwing with flow form screws

In addition to the plastic side, the assembly of a complex lightweight system in mixed construction can also be carried out using metal-side direct screw methods. In body-in-white construction, joining techniques using flow drilling screws have become particularly well established. These locally exploit the plastic flow properties of the metal component, such as the EJOT FDS® screw system, which offers significant advantages when joining chemically incompatible materials, see e.g. [35, 36]. Analogous to the plastic-side direct screwing, considerable transverse forces are generated at the joints of textile-reinforced thermoplastics during the production of FDS® joints. These induce a high-notch stress concentration due to the high screw-in torques during setting and the non-material-specific design of the screw head geometry, which leads to high interlaminar shear stress peaks [37]. For highly stressed mixed structures with thin-walled plastic components, direct bolting with flow-hole forming screws is therefore unsuitable.

2.3 Adhesive joints

Adhesive joints have been successfully used in vehicle construction for a long time and have proven their worth for mixed connections with metal and plastic components, see [38]. Adhesives are used in particular in the manufacture of front ends and add-on parts such as doors, tailgates, and engine hoods, e.g. [39]. An adhesive is a non-metallic material that joins parts by surface adhesion and internal strength (cohesion). Adhesive bonds can also be used to connect chemically incompatible materials, such as metals and plastics, in a quasi-substance-bonding manner. In this connection, the loads are introduced over a large area and almost homogeneously into the components of the mixed construction system. In contrast to bolted, riveted, and screwed joints, the fiber architecture on the composite component is not affected by the joining process, so that bonded joints for FRP components are generally considered to be particularly suitable for the material, see e.g. [40]. Adhesive bonding technology can be used to produce multifunctional joining systems which, in addition to force transmission, also perform sealing and vibration damping functions. In contrast, the high cost of pretreating the parts to be joined, such as activating the joining surfaces by means of radiation treatment, etching, grinding or degreasing with hydrocarbon compounds, to ensure reproducible joining properties—especially in large-scale production—is very cost-intensive [41].

2.4 Thermal contact joining

Thermal contact joining of short glass fiber-reinforced polyamide (GF20-PA6) and high-alloy austenitic steel X5CrNi18-10 (1.4301) is described in [42]. Prior to the joining process, pins are welded onto the metal component using modified gas metal arc welding (GMAW) technology. In contrast to conventional GMAW welding processes, the wire electrode consisting of the joining component material is not completely melted but is placed in a defined manner on a metal sheet so that the head geometry and length of the pins can be varied. The joining process is then carried out by pressing the overlappingly positioned joining parts. For this purpose, the FRP component is first plasticized up to the melting temperature of the matrix, followed by pressing the joining partners. Depending on the type and degree of mechanical surface activation, the plastic melt flows into the unevenness of the metal surface, resulting in form-fit and material-fit bonding mechanisms. The strength of the mixed joint is primarily determined by the pins of the metal component penetrating the FRP during the joining process.

2.5 Further joining methods for FRP

Further joining methods for FRP that are state of the art and/or the subject of current research work are summarized in the following.

Welding of metal/plastic-based mixed structures was investigated e.g. in Refs. [15, 16, 43–46], where primarily ultrasonic and induction welding processes were considered. In this process, adhesion is based on the penetration of the polymer into the rough surface structure of the metal, the formation of dispersion forces and hydrogen bridge bonds, and the formation of weak covalent bonds between the metallic joining partner and the functional groups of the polymeric components. For the wetting of the joining partners, the specific surface and interfacial energies play a decisive role, which in this context is also referred to as thermodynamic adhesion [47]. Joining by ultrasonics of continuous fiber-reinforced organic sheets with metals was considered, for example, in Refs. [14, 48]. It was found that the metal component penetrates locally into the thermoplastic matrix of the composite component due to the ultrasonic vibrations acting perpendicularly to the joining zone, so that the metal component can enclose the exposed reinforcing fibers in the ductile state. Comparatively high-bond strengths are achieved by embracing the load-bearing fibers. In these investigations, mainly good formable metals such as Al99.5 were used. The induction welding process requires the electrical conductivity of the joined parts. Due to the induced resistance effects and the magnetic hysteresis during dissipation, heat is generated which leads to the plasticization of the thermoplastic matrix of the FRP component. The subsequent pressing allows the parts to be joined by a quasi-material bond, analogous to adhesive bonding, see [46, 47, 49].

No additional joining element is required for collar joining [3, 50, 51]. To produce the mixed joint, a hole is first drilled in the metal sheet. A collar is then formed from the metal sheet at this point with the aid of a die. In the process, a groove for the formation of an undercut on the collar of the metal sheet is created by the previous embossing with a matrix. In the actual joining process, the metal component with the collar is pressed into the plastic part, so that the undercut creates a form-fit connection between the joining partners. Collar joining is not recommended for continuously reinforced thermoplastic composite components because the reinforcing fibers are severed at the joint during pressing.

In the case of injection riveting, the connections are implemented using an injection molding technology adapted for this purpose [52]. Disadvantages of the technique, similar to stamp riveting [53], are the necessary two-sided accessibility as well as the complex plant and process technology.

2.6 Machining of FRP components

When drilling FRP, delamination and microcrack are often caused at the bore edges during the entry and exit of the drilling tools. The degree of damage is strongly influenced by the condition of the cutting blades, see [19]. At the exit point, the delamination is mostly caused by the abrupt penetration of the drill. In Refs. [54–56] it is recommended to support the FRP workpiece at the exit point by appropriate shims with softer materials, such as wood or aluminum plates, and to mount it with low vibration. In addition, the drilling process for thermoplastic fiber composites often leads to decomposition of the polymer matrix. To a limited extent, this process can be counteracted with cooled tools and reduced cutting and feed rates, see e.g. [57]. Damage to the bore edge can be largely avoided with circular or orbital milling processes [19]. To create a through hole, the drilling tool is fed along a helical path to the FRP workpiece. Thus, in contrast to conventional drilling technology, boreholes with different diameters can be produced with the aid of one tool. Furthermore, the axial machining forces and the associated tool wear are comparatively low if the cutting values are selected appropriately. The production of internal threads is described in Refs. [19, 23, 58] for glass fiber-reinforced plastic (GFRP) and CFRP components. Internal threads can be produced in FRP components by tapping, cutting, milling and thread forming. In thread forming, the thread flanks are shaped by a forming process.

For this purpose, the forming studs of the thread forming tool are pressed into the pre-drilled workpiece, whereby the thread flanks are formed due to plastic deformation. In contrast, drilling and milling processes represent material-removing machining techniques. In tapping, the thread flanks and thread tips are already significantly damaged when the core hole is inserted at the base of the thread. The resulting spalling, fiber bundle fractures and kinks as well as distortions often considerably reduce the strength behavior of the screwed connection. Thread milling, on the other hand, causes much less damage with comparatively low machining forces.

3. Design of through holes suitable for FRP

The load-bearing fibers of the highly anisotropic FRP must be oriented around through-holes in accordance with the load path in order to utilize the full material potential. The analysis of construction principles in nature and their adaptation to disturbance points provides guidance in optimizing the shape of these areas. For example, in trees, the maintenance of a constant stress distribution at the disturbance zone is observed, see e.g. [59, 60]. If a tree is damaged, e.g. by cracks, decay or fractures, the resulting stress increases are reduced or absorbed by the accumulation of additional or firmer material near the defect [61, 62] (**Figure 1a**).

Since wood has maximum strength and stiffness properties only in the direction of the fibers, the tree always strives to align its wood fibers along the highest stress, see for example [63]. The orientations of the wood fibers thereby largely correspond to the direction of the greatest principal stress, reducing the shear between the fibers and increasing the material utilization. This design principle of nature can be used in technical FRP applications to increase strength, especially at notch-sensitive component sections (e.g., at holes or cutouts), as a design model for the load-appropriate arrangement of reinforcing fibers at load application areas. Based on this principle, optimization algorithms have been developed in Refs. [64, 65] to align the fibers in the direction of the principal stress trajectories in a load-appropriate manner (**Figure 1b–d**). The methods presume freedom from shear stresses in the principal stress system. Since the fiber composite component has the largest modulus of elasticity in the longitudinal direction of the fibers, the FRP component is stiffness-optimized by the specific arrangement of the reinforcing fibers in the principal stress direction.

In contrast to thermoset FRP, where the redirection of the fibers at the load application points must already be taken into account in the preform process (**Figure 2a**), the reorientation of the reinforcing fibers in textile-reinforced thermoplastics can be

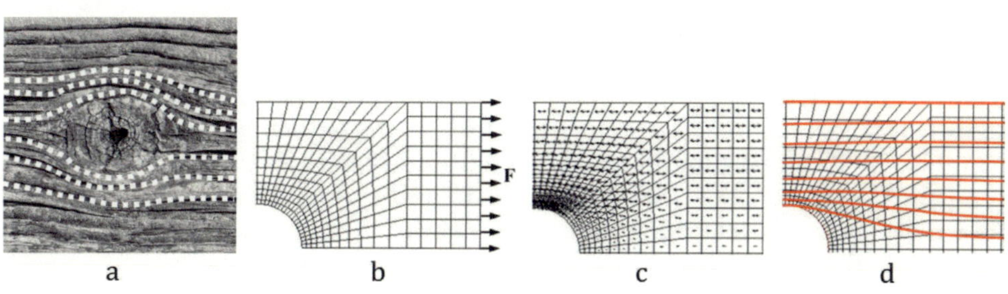

Figure 1.
(a) Fiber orientation around a knothole, (b) numerical model, (c) principal stress trajectories, (d) optimized fiber orientation [5].

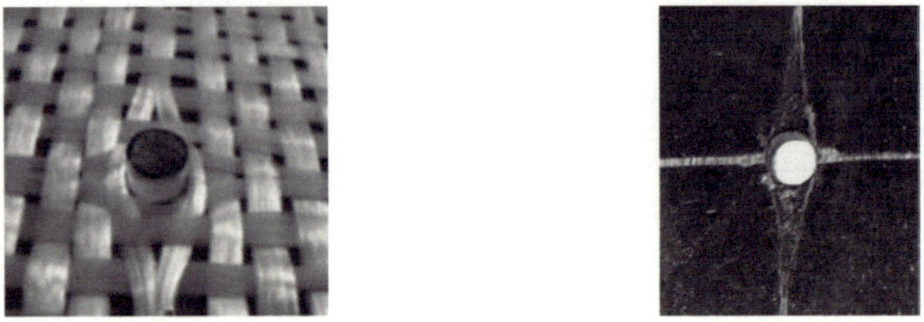

Figure 2.
Reorientation of the fibers (a) of a textile fabric around an insert prior to the resin insertion and (b) in a thermoplastic FRP [5].

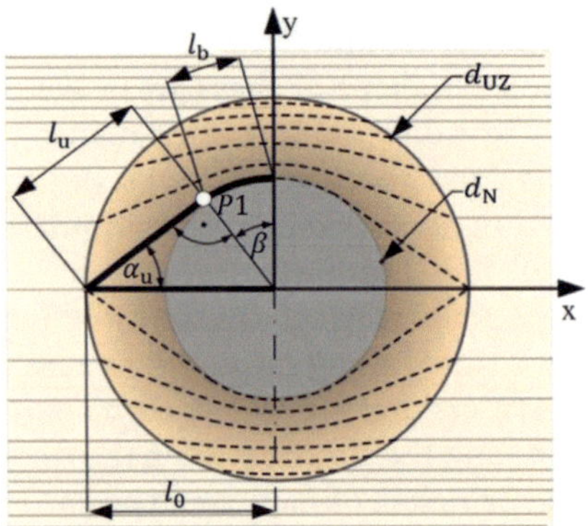

d_{UZ} diameter of the reforming zone
d_N diameter of the needle
l_0 length of the unstressed fiber
$P1$ tangent intersection point
l_u length of reoriented fiber to P1
l_b arc length
α_u reorientation angle
β wrap angle at the needle edge

Figure 3.
Schematic representation of fiber redirection on a unidirectional prepreg layer [5].

carried out subsequently on the component, see e.g. [66–70]. To ensure radial displacement of the fibers within the matrix (**Figure 2b**), a defined area must first be plasticized at the load introduction area of the organic sheets. The viscosity of the matrix can be reduced in a targeted manner by local heat input, so that the reinforcing fibers can be reoriented in the forming zone, for example with the aid of a needle [5].

Due to the local heating, however, the fibers remain firmly clamped in the adjacent colder matrix material and are therefore stretched differently during reorientation with the needle tool. Thus, in addition to the diameter of the through hole, the permissible strain at break of the fiber component in particular must be given special consideration when dimensioning the reforming area. In the following model (**Figure 3**), the most deflected and most elongated fibers of a unidirectional single layer at the needle edge are considered [5].

Since the diameter d_N of the needle is given and the reoriented fiber with length l_u is tangential to the bore edge at point $P1$, the reorientation angle and the wrap angle β are equal. Since the maximum fiber strain is also known as a material parameter, the transcendental Eq. (1) can be derived and solved numerically. In this way, Eq. (2) is used to determine the diameter of the reforming zone. A detailed derivation of the equations and evaluation examples is given in [5].

$$\cos\alpha_u + \alpha_u \cdot \sin\alpha_u - 1 = \varepsilon_B \tag{1}$$

$$d_{UZ} = \frac{d_N}{\sin \alpha_u} \tag{2}$$

The filaments located directly at the bore edges are prestressed in the range of failure strain. To ensure that further loading does not lead to fiber breaks, the heating region must be selected larger than the diameter d_{UZ} of the reforming zone [5].

In Ref. [5], a test rig was developed and implemented. The results were validated in bolt tensile tests following DIN 65562. Reference specimens were produced with spiral and wood drills. Especially with the spiral drill, delamination is caused at the bore edges. The test results of Ref. [5] show that the utilization of the high specific strength and stiffness properties of organic sheets can be increased by more than 28% by designing the load application point to suit the load. In addition, the deformation work of the specimens with the hot-formed holes is 25% higher than that of the reference specimens.

Different failure mechanisms can be observed in the breakage images and fracture patterns of the tested components. Whereas machined specimens fail primarily by strong delamination, cracks and intermediate fiber fractures at the edge of the hole, the test specimens with formed holes only show fiber fractures at the 90° layers arranged transverse to the loading direction [5].

In contrast to the spiral-drilled specimens, the strength $\sigma_{L,2\%}$ at a hole widening of 2% according to DIN 65562 is lower for the hot-formed specimens. Although the force flow in drilled specimens is interrupted at the load application point, the 0° plies before micro buckling of the reinforcing fibers still fully contribute (**Figure 4**), so that increased strength and stiffness values are the result in the initial stage. This is not the case with organic sheets with hot-formed holes, since from the beginning the 90° plies (with very low stiffness) running transverse to the loading direction primarily take up the load. In contrast, only very low axial loads can initially be introduced into the 0° layers via the flexible matrix. Overall, hot-formed holes, therefore, exhibit softer behavior compared to holes drilled by machining, resulting in a lower $\sigma_{L,2\%}$ strength. With an increase in load, however, the load-bearing capacity rises due to targeted stress redistribution and ultimately results in a significantly higher hole bearing strength compared to drilled specimens.

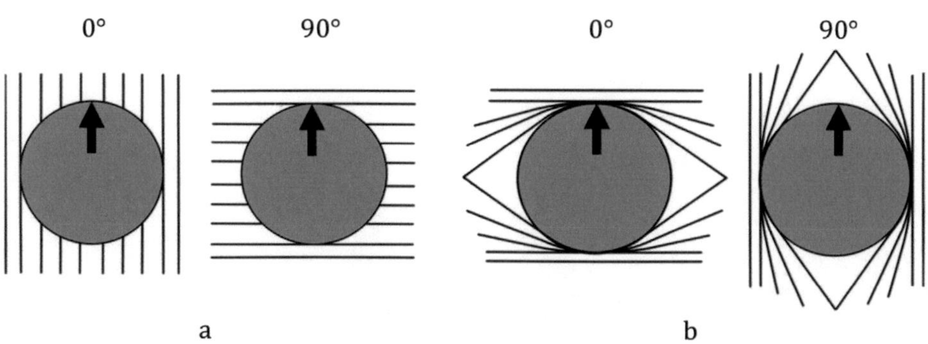

Figure 4.
Loading of the 0° and 90° layers in orthogonal multilayer composites with (a) machined holes and (b) hot-formed holes [5].

4. Thermomechanical flow drill joining

In Ref. [5], comprehensive investigations were carried out on the manufacturing of metallic bushings by means of the flow drilling process as well as on their quality evaluation. This process uses a rotating hard metal mandrel which penetrates into the metal component (**Figure 5**). As a result of the frictional heat generated, the resistance of the base material is reduced to such an extent that the drill mandrel can be pushed through the metal component with little force. As soon as the metal plasticizes at the entry point of the working area, it initially begins to flow at the taper in the opposite direction to the feed direction. At the same time, as the sheet is pushed through, the material is dragged along in the feed direction as a result of combined friction/flow processes, causing a bushing to form on the underside.

Based on this technology, thermomechanical flow drill joining for hybrid components made of FRP and metal, which is also suitable for economical large-scale production, was developed. The process is characterized by short process times, force-flow-compatible and high-strength joining properties, and a high degree of lightweight design. In contrast to many common processes, no auxiliary joining element is required. To produce a FDJ joint, the plastic flow properties of the metal component are specifically used. A rotating mandrel is used to thermomechanically form a bush from the metallic base material as described above. The heat generated causes the thermoplastic to plasticize. Alternatively, further heat can be introduced by heating elements. During the forming process, the bush is pushed directly through the thermoplastic FRP component (**Figure 6a**), which is also plasticized locally and then turned over positively when the mandrel is reset or in a further process step (**Figure 6b**). The fibers are reoriented around the flat joints (**Figure 6c** and **d**).

In Ref. [5], a joining device was developed, designed and implemented. It provides the industry with an efficient technology for the production of material-specific joints between FRP and metals using FDJ technology. In a comprehensive parameter study, the optimal process variables for joining FRP of different thicknesses and different metal alloys were determined. The comparison of the head and shear tensile strengths based on DIN EN ISO 14272 and DIN EN ISO 14273 of the new joining technology with conventional methods such as riveting and flow-forming screws shows a significantly higher load-bearing capacity of the FDJ joints (**Figure 7**). The process parameters with which the specimens for the tests in **Figure 7** were produced are given in Ref. [5] page 79 Figures 7–13.

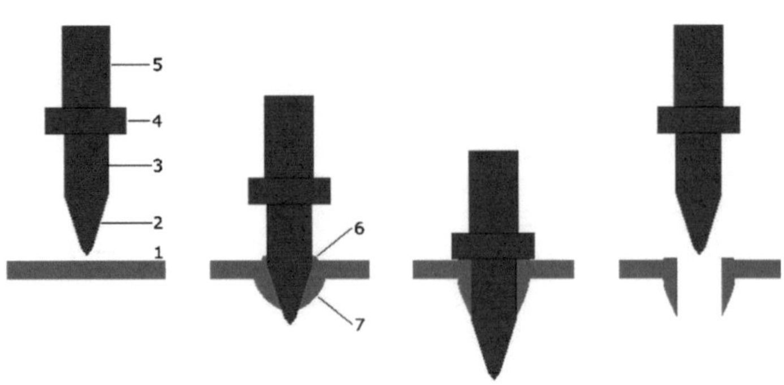

Figure 5.
Flow drilling method with 1 working area, 2 cone, 3 cylindrical area, 4 collar, 5 shank, 6 material flow against feed direction, 7 bush, according to ref. [71].

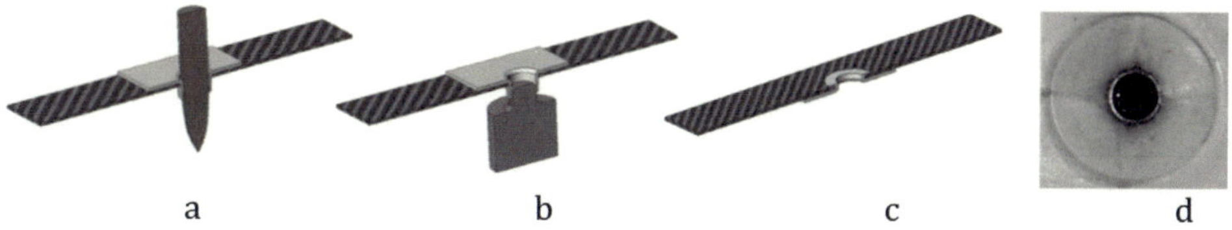

Figure 6.
Principle for manufacturing a FDJ connection [5].

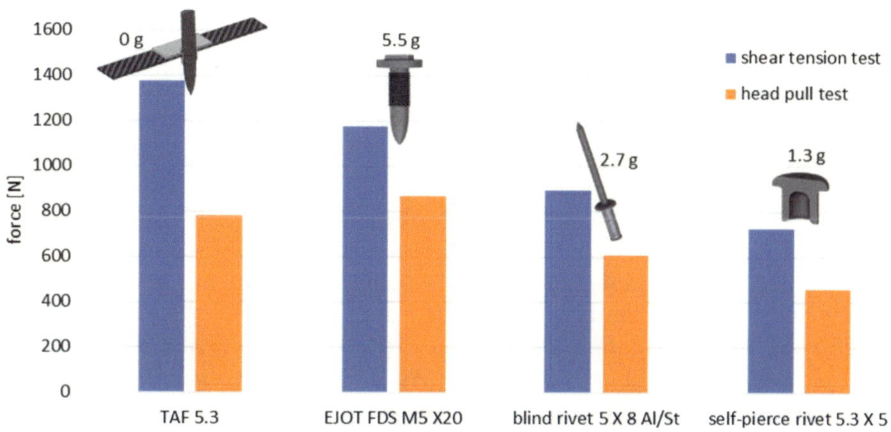

Figure 7.
Comparison of head and shear tensile strength of FDJ joints with conventional joining methods [5].

With the developed failure hypothesis for the sizing of the joints with the FEM, the method also meets the requirements of the industry for the calculation and dimensioning of the joined components [5].

5. CMT-pin technique

The FDJ technology can be used to create highly stressable material-compatible joints between thermoplastic FRP and metals. However, many applications, e.g. in automotive engineering, require a closed joint, which is not provided by the FDJ process due to the remaining through-hole.

Therefore, joining of FRP with metals by cold metal transfer pin welding (CMT pin) is proposed as an alternative automatable process with one-sided accessibility. For improved low-spatter droplet separation with minimal heat input compared to MIG and MAG welding, this arc welding process uses computer controlled backward movement of the wire during short-circuiting, followed by movement toward the workpiece. The technology was developed by the company Fronius International GmbH and can be adapted for welding pins. Fronius provides appropriate operating curves for this purpose.

Sequential joining, in which the pins are first welded and then the polymer component is pressed on or, in the case of duroplastic systems, cast on, is well known. In the process described below, the FRP component is joined in situ.

In the first process step, the FRP is positioned over the metal and heated, e.g. by means of an infrared radiator, until the thermoplastic matrix softens (**Figure 8a**). Subsequently, the welding wire is pushed through the organic sheet by feeding the wire or a movement of the robot and the fibers are displaced according to the load path (**Figure 8b**). The use of an endless welding wire as joining element (pin) ensures flexible pin lengths even on graded wall thicknesses of the FRP. Immediately after contacting the wire, it is welded to the metal. The so-called pin foot is formed, the shape of which can be optimized by moving the wire forwards or backwards. Finally, the wire is separated by a separation pulse (**Figure 8c**). The head shape of the pin can be adjusted by selective adaptation of the process parameters. The connection is form-fit. The pin shaft takes up the forces in the plane and the head orthogonally to it.

In Refs. [72, 73], a heating strategy using infrared radiation was developed. The heating of flat sheets of carbon fiber-reinforced polyamide 6 (CF-PA6) with infrared radiation was modeled using the finite element method (FEM), allowing the temperature distribution in the sample to be studied in greater detail than is possible experimentally. The infrared radiation heats the upper surface of the CFRP sheet. Some of this heat is released back into the environment by convection and by radiation. In the plate, heat transport takes place by means of conduction. Due to the time dependence of the heating process, the heat capacity and density of the material are needed in the simulation. The emission coefficient describes the heat transfer by radiation and the heat transfer coefficient describes the convective heat transfer. Both values were determined in a parameter study. For this purpose, the values are varied systematically and the FEM calculations are performed in each case. The temperatures as a function of time were compared with the experimental data at several points along the plate diagonals. The density of the PA6-CF is 1.434 g/cm^3. Heat capacity (**Figure 9**) was measured using differential scanning calorimetry technique with a heating rate of 10 K/min. The thermal conductivity calculated according to Refs. [11, 74–76] is 4.72 W/(m•K) in the fiber direction and 0.48 W/(m•K) perpendicularly.

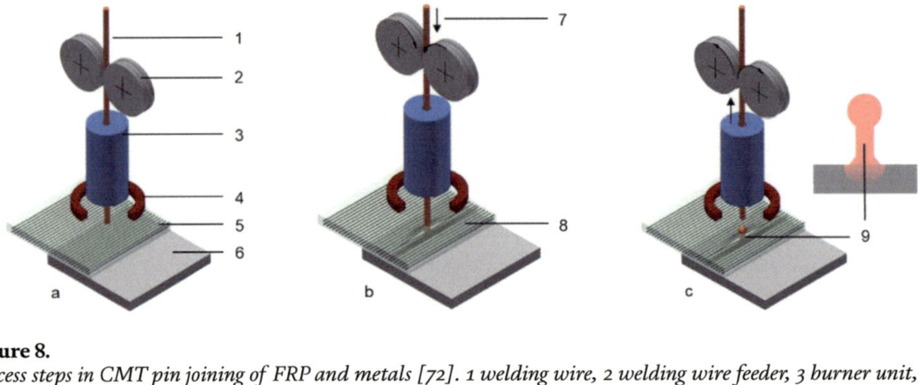

Figure 8.
Process steps in CMT pin joining of FRP and metals [72]. 1 welding wire, 2 welding wire feeder, 3 burner unit, 4 infrared heater, 5 FRP, 6 steel part, 7 feeding direction of the wire, 8 reoriented fibers, 9 pin with head.

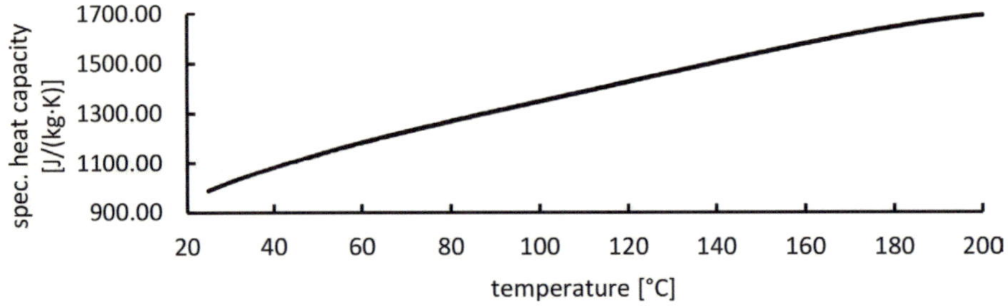

Figure 9.
Experimentally determined temperature-dependent heat capacity of the CFRP investigated [72, 73].

A layup with a sequence of 0 and 90° plies was investigated. The plate thicknesses considered are 2 and 3.8 mm. The calculations were performed with the software Abaqus 2019. To validate the FEM models, the heating of square CFRP sheets with an edge length of 150 mm at different distances of the IR radiator to the CFRP surface was studied experimentally. To observe the area distribution of the temperature on the irradiated side, 16 thermocouples were pressed into the CFRP sheet along the diagonals at different distances from the center of the sheet. The 250 W Omega radiator is assumed to be a circular radiator in Refs. [72, 73] and the model is built with a cyclic symmetry. Since losses are assumed, the radiator power is assumed to be 210 W in the model. The smallest deviation between experiment and FEM was obtained with

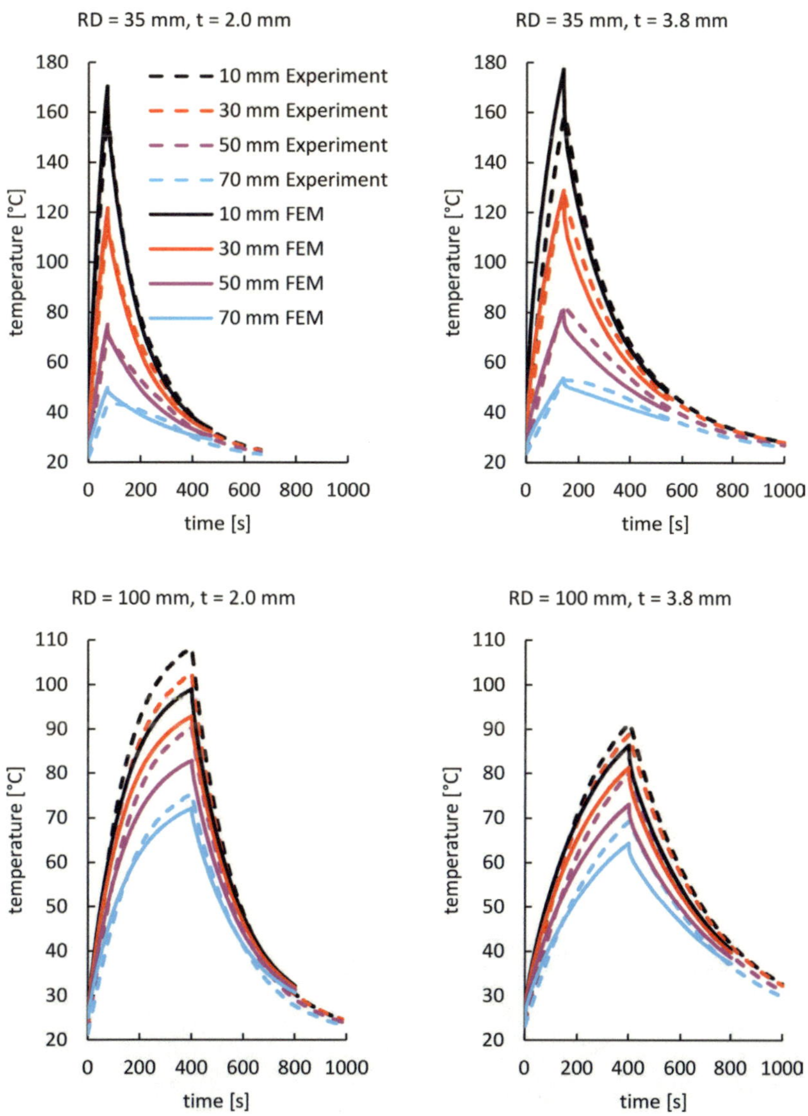

Figure 10.
Comparison of the temperature curves on the top surface between experiment and 3D simulation for different distances of the measuring points to the center of the plate along the diagonal [72, 73]. RD radiator distance, t thickness of the CFRP.

a constant heat transfer coefficient of 2.5 W/(m^2•K) and an emission coefficient for the CFRP of 0.95 and for the radiator of 0.6 (**Figure 10**).

The comparison of the temperature fields recorded in the experiment with a thermal imaging camera with the simulation results also indicates a high level of agreement.

With the help of the models built in this way, it was possible to analyze different radiators and process cycles in Refs. [72, 77] and thus develop a heating strategy suitable for CMT pin joining.

On this basis, comprehensive parameter studies were carried out in Refs. [72, 77] on the joining of CF-PA6 with sheets of 1.5528 with a 150 g/m^2 Al-Si coating. It was found that the electrically conductive CFRP cannot be joined by this method. The welding wire penetrates the CFRP, but moves back to the surface of the organic sheet to ignite the arc. When the arc is ignited, the CFRP is severely damaged due to the strong and rapid heat input. Since the process curves are pre-programmed by the manufacturer, it was not possible to develop a welding strategy in Ref. [72] in which the arc ignites on the metal. Therefore, the joining of dielectric glass fiber-reinforced polyamide 6 (GF-PA6) was investigated as an alternative. With this material, the arc ignites on the steel sheet as intended. Even with a distance of 1.5 mm between the steel and the GFRP, some of the fibers are not completely diverted around the pin but are pushed out downwards, so that they disrupt the welding process. The reason for this is that with the corresponding welding parameters, not only the pins have spherical heads, but also the welding wire. In a further parameter study, a two-stage process sequence was developed in which the heated and thus softened organic sheet is first penetrated by the robot movement with a pointed welding wire, so that all joints are pre-holed according to the material and then the pins with spherical heads are welded. By means of microcomputer tomographic analyses, the deflection of the fibers around the pin could be demonstrated (**Figure 11**). Micrographs of the weld metal show complete bonding of the pin foot to the base material (**Figure 12**).

Using the process described, in which the locally preheated organic sheet is first penetrated at the joints with a pointed welding rod and then the pins are welded, it was possible to implement functional demonstrators in Refs. [72, 77] as GFRP sheets welded onto steel-hat profiles. This represents thrust fields, e.g. in vehicle construction. A comparison with glued specimens was made in a three-point bending test. An 8% higher flexural strength and an approx. 25% greater work of fracture of the CMT pin joined specimens can be determined [72, 77].

For detailed analyses of this joining method and validation in experimental tests, see Refs. [72, 73, 77].

Figure 11.
Microcomputer tomographic analysis of GFRP sheet after penetration [72, 77].

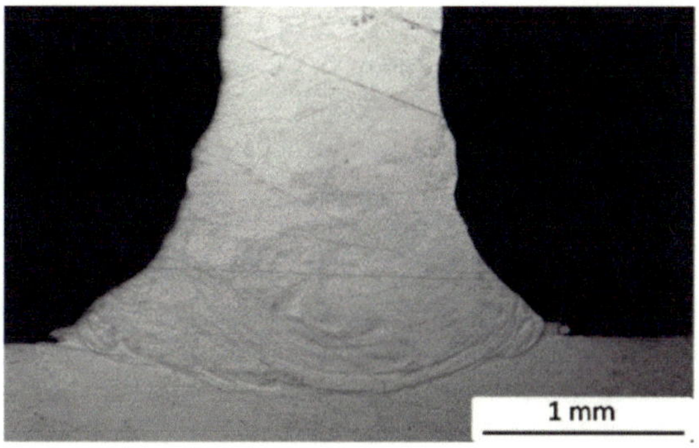

Figure 12.
Micrograph of pin foot [72, 77].

6. Conclusion and outlook

It has been shown that conventional methods such as screwing, riveting, and bolting are unsuitable, especially for joining continuous fiber-reinforced FRP with metals, since the load-bearing fibers are cut during the pre-hole cutting operations or when the joining elements are set. This leads to considerable notch stress concentrations at the joints. To reduce these, a method based on nature's construction methods was presented in which through-holes are created in thermoplastic FRP by reorienting the fibers in this area to match the load path around the point of disruption. To do this, the polymer matrix is softened locally by applying heat and penetrated with a needle or mandrel. Based on the strain at break of the fibers, the diameter of the forming area is calculated.

Based on these results, flow hole forming was used as a starting point for the development of a joining technology for connecting FRP and metals in a way that is suitable for the material: thermomechanical flow drill joining. In this process, a mandrel forms a bush from the metal component. The heat generated in the process softens the plastic of the organic sheet so that the fibers are displaced instead of severed. Another joining process that can be fully automated is CMT pin welding. In this process, the matrix is also softened by the application of heat, e.g. by infrared

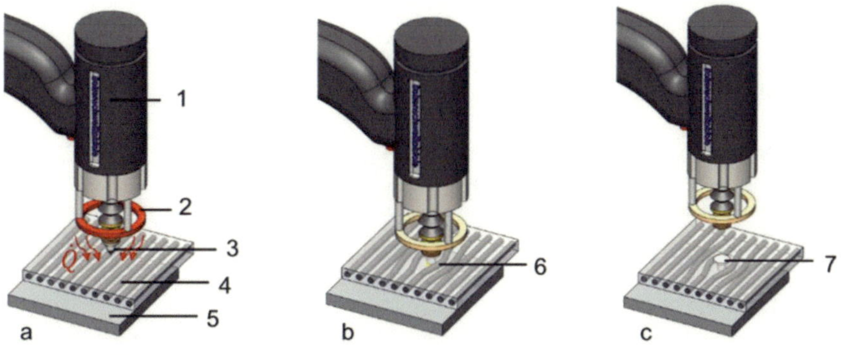

Figure 13.
Stud welding process for joining thermoplastic FRP and metals. 1 stud welding gun, 2 heater, 3 stud, 4 FRP, 5 steel part, 6 reoriented fibers, 7 stud joint.

radiation. The plastic component is then penetrated with the welding wire, displacing the fibers in the joining area and realigning them to suit the load path. A separation pulse is finally used to form a head, thus securing the organic sheet against lifting from the metal component. The advantage of the CMT pin technique over the FDJ process is that the joints are closed, but they are not flat due to the pin heads. Furthermore, CMT pin welding can currently not be used to join electrically conductive CFRP with the manufacturer's predefined process curves.

As an alternative to these two technologies, the research group Sustainable Lightweight Technologies ZenaLeb of the Fraunhofer Institute for Applied Polymer Research IAP Division Polymeric Materials and Composites PYCO and the Chair of Polymer-based Lightweight Design of the Brandenburg University of Technology Cottbus-Senftenberg in Germany is currently analyzing the capability of a new method based on stud welding [78]. In this process, the FRP is pre-heated (**Figure 13a**) analogously to the CMT pin technique. It is then penetrated by a joining element, a stud in the form of a nail, so that the fibers are reoriented according to the load path (**Figure 13b**). Due to the adjustable stud geometry with a flat head, in contrast to the CMT pin technique, plane joints (**Figure 13c**) can even be realized with CFRP, which, in contrast to the FDJ process, are closed.

Acknowledgements

The results presented in this chapter were generated in several research projects funded by the Federal Ministry for Economic Affairs and Climate Action and the Federal Ministry of Education and Research.

Author details

Holger Seidlitz[1,2,3*], Lars Ulke-Winter[1,2,3], Lucas Ost[1,2,3] and Felix Kuke[1,2,3]

1 Chair of Polymer-Based Lightweight Design (PbL), Brandenburg University of Technology Cottbus-Senftenberg (BTU), Cottbus, Germany

2 Fraunhofer Institute for Applied Polymer Research IAP Research Division Polymeric Materials and Composites PYCO, Wildau, Germany

3 Center for Sustainable Lightweight Technologies ZenaLeb of the Fraunhofer IAP, Research Division PYCO, Cottbus, Germany

*Address all correspondence to: fg-leichtbau@b-tu.de

References

[1] Sauer M, Schüppel D. Composites Market Report 2021 - the Global Market for Carbon Fibers and Carbon Composites. Freely Accessible Short Version. Berlin: Composites United e.V; 2022

[2] Ehrenstein GW. Handbuch Kunststoff-Verbindungstechnik. München: Hanser Verlag; 2004

[3] Endemann U, Glaser S, Völker M. Verbindungstechnik für Kunststoff-Metall-Hybridstrukturen: Kunststoff und Metall im festen Verbund. Ludwigshafen, Germany: Kunststoffe; 2002. pp. 110-113

[4] Neitzel M, Mitschang P, Breuer U. Handbuch Verbundwerkstoffe - Werkstoffe, Verarbeitung, Anwendung. 2nd ed. München: Hanser Verlag; 2014

[5] Seidlitz H. Entwicklung von kraftflussgerechten Verbindungstechniken für Mischbauweisen mit thermoplastischen Faserverbunden und Metallen. In: Dissertation. Chemnitz: Technischen Universität Chemnitz, Fakultät für Maschinenbau; 2013

[6] Ruther M, Jost R, Freitag V, Peitz V, Piccolo S, Brüdgam S, et al. Fügesystemoptimierung zur Herstellung von Mischbauweisen aus Kombinationen der Werkstoffe Stahl, Aluminium, Magnesium und Kunststoff. Final Report. Paderborn: Universität Paderborn, Werkstoff- und Fügetechnik; 2003

[7] Hahn O, Bye B, Draht T, Lübbers R, Ruther M, Zilg C, et al. Fügen von faserverstärkten Kunststoffen im strukturellen Leichtbau. Paderborn: Universität Paderborn; 2004

[8] Lesemann M, Sahr C, Hart S, Taylor R. SuperLIGHT-CAR - the multi material car body. In: 7. LS-DYNA Anwenderforum. Bamberg: RWTH Aachen University, Institut für Kraftfahrzeuge; 2008

[9] Steinig M. Hochgeschwindigkeitsbolzensetzen für hochfeste und hybride Strukturverbindungen. Tagungsband T32 des 31. EFB-Kolloquiums Blechverarbeitung. Bad Boll: EFB: Hochfeste und hybride Materialien - Schnelle Umform- und Fügeverfahren; 2011

[10] Schöll R, Friedrich HE, Erdl G. Innovative Fahrzeugleichtbaustruktur in Spant- und Space-Frame-Bauweise. Zeitschrift für Produktentwicklung und Ingenieur-Werkstoffe. 2009:37-38

[11] Schürmann H. Konstruieren mit Faser-Kunststoff-Verbunden Berlin. Heidelberg: Springer-Verlag; 2007

[12] Moser K. Faser-Kunststoff-Verbund: Entwurfs- und Berechnungsgrundlagen. Düsseldorf: VDI-Verlag; 1992

[13] Krüger S, Wagner G, Eifler D. In: Degischer HP, editor. Eignung des Metall-Ultraschallschweißen zum Fügen von Verbundwerkstoffen. Weinheim: WILEY-VCH; 2003

[14] Ageorges C, Ye L, Hou M. Advances in fusion bonding techniques for joining thermoplastic matrix composites: A review. Composites Part A: Applied Science and Manufacturing. 2001;**32**(6):839-857

[15] Balle F, Huxhold S, Emrich S, Wagner G, Kopnarski M, Eifler D. Influence of heat treatments on the mechanical properties of ultrasonic welded AA 2024/CF-PA66-joints.

Advanced Engineering Materials. 2013;**15**:837-845

[16] Balle F, Wagner G, Eifler D. Ultrasonic metal welding of aluminium sheets to carbon fibre reinforced thermoplastic composites. Advanced Engineering Materials. 2009;**8**:35-39

[17] Ehrenstein GW. Mit Kunststoffen konstruieren. Universität Erlangen-Nürnberg. München: Hanser-Verlag; 2020

[18] Bergmann HW. Konstruktionsgrundlagen für Faserverbundbauteile Berlin. Heidelberg: Springer-Verlag; 1992

[19] Weinert K. Bearbeitung von Textilverbunden. In: Hufenbach W, editor. Textile Verbundbauweisen und Fertigungstechnologien für Leichtbaustrukturen des Maschinen- und Fahrzeugbaus. Dresden: SDV-Die Medien AG; 2007. pp. 90-99

[20] Niklewicz J, Ferris DH, Nunn GJ, Sims GD. The use of pin bearing data for the preliminary design of 'bolted' joints. In: International Conference on Joining and Repair of Plastics and Composites. London: Woolstencroft, D. H.; 1999. pp. 13-22

[21] Hufenbach W, Gude M, Freund A. Simulation des Schädigungsverhaltens von thermomechanisch beanspruchten Blindnietverbindungen in CFK/Al-Strukturen. NAFEMS Magazin. 2008;(10):30

[22] Kroll L, Mueller S, Mauermann R, Gruetzner R. Strength of self-piercing rivited joints for CFRP/aluminium sheets. In: 18th International Conference on Composite Materials (ICCM18). Jeju Island, Korea; 2011

[23] Weiner K, Kempmann C. Gewindefertigung in faserverstärkten Kunststoffen. Kunststoffe. 2004;**7**:44-48

[24] Renz R, Ahlers-Hestermann G, Humm H. Mechanisches Fügen von Kunststoffen - Direktverschraubung von Formteilen aus Kunststoffen und Gewindeeinsätze Teil 1. Joining Plastics. 2008;**2**(1):64-69

[25] Küting J, Hahn O. Entwicklung des Fliessformschraubens ohne Vorlochen für Leichtbauwerkstoffe im Fahrzeugbau. In: Dissertation. Paderborn: Universität Paderborn, Laboratorium für Werkstoff- und Fügetechnik; 2004

[26] Tome A, Dratschmidt F, Ehrenstein GW. Direktverschraubungen in Kunststoff - Günstig auf lange Sicht? Plastverarbeiter. 1998;**49**:124-127

[27] Küting J, Meschut G. Direktverschrauben ohne Vorlochen zur Realisierung einseitig zugänglicher Verbindungen im Karosseriebau. Blech Rohre Profile. 2003;**2**:33

[28] Dratschmidt F. Zur Verbindungstechnik von glasfaserverstärktem Polyamid - Schrauben und Inserts. In: Dissertation. Erlangen: Universität Erlangen-Nürnberg, Lehrst. f. Kunststofftechn; 1999

[29] Tome A. Vorspannkraftrelaxation von Kunststoff-Direktverschraubungen. In: Dissertation. Erlangen: Universität Erlangen-Nürnberg, Lehrst. f. Kunststofftechn; 2000

[30] Pyper M. Kunststoffteile direkt verschraubt. Konstruktion & Engineering. 2006;**9**:80-81

[31] Renz R, Ahlers-Hestermann G, Humm H. Mechanisches Fügen von Kunststoffen - Direktverschraubung von Formteilen aus Kunststoffen und Gewindeeinsätze Teil 2. Joining Plastics. 2008;**2**(2):142-148

[32] Kerb-Konus-Vertriebs-GmbH. Gewindeeinsätze für Kunststoffe und Holz: Produktdatenblatt Nr. 30. 2012. Available from: http://www.kerbkonus.de/ [Accessed: 2012; Produktdatenblatt Nr. 30]

[33] Kerb-Konus-Vertriebs-GmbH. Gewinde für dünne Formteile: Produktdatenblatt Nr. 40. 2012. Available from: http://www.kerbkonus.de/

[34] Wilhelm Böllhoff GmbH & Co. KG. Hochbelastbare Gewindeeinsätze zum Umspritzen: Produktdatenblatt. 2012. Available from: http://www.boellhoff.de/

[35] Küting J. Entwicklung des Fließformschraubens ohne Vorlochen für Leichtbauwerkstoffe im Fahrzeugbau. In: Dissertation. Paderborn: Universität Paderborn, Laboratorium für Werkstoff- und Fügetechnik; 2004

[36] Quitter D. Leichtbau durch Mischbauweise, neue Strukturkonzepte und angepasste Fügetechniken. Würzburg: Konstruktionspraxis; 2007

[37] Seidlitz H, Kroll L, Ulke-Winter L. Kraftflussgerechte Verbindungstechniken für Hochleistungskomponenten in Mischbauweise. In: euroLITE - Internationale Fachmesse für Leichtbaukonstruktion. Salzburg; 2009

[38] Rasche M. Kleben im Fahrzeugbau. Adhäsion. 1987;(9):13-23

[39] Maurer B, Schäfers C, Maciej M. Kleben und Dichten von Bördelfalznähten im Karosseriebau. In: 13. Paderborner Symposium Fügetechnik. Paderborn; 2006

[40] Ridzewski J. Konzepte zur Gestaltung rohrförmiger CFK-Aluminium-Hybrid-Verbunde. In: Dissertation. Clausthal: TU Clausthal; 2005

[41] SeidlitzH.Grundlagenuntersuchungen zur Kennwertermittlung, Berechnung, Modellierung und Dimensionierung von Dickschicht-Klebverbindungen. Diploma Thesis. Cottbus: HS Lausitz; 2005

[42] Nestler D, Bayreuther V, Trautmann M, Wielage B, Hälsig A, Podlesak H, et al. Wärmekontaktfügen von Stahl/CFK-Werkstoffverbunden. In: Verbundwerkstoffe und Werkstoffverbunde. In 18. Symposium Verbundwerkstoffe und Werkstoffverbunde. Chemnitz; 2011. pp. 346-353

[43] Balle F, Wagner G, Eifler D. Ultrasonic spot welding of aluminum sheet/carbon fiber reinforced polymer – Joints. Materialwissenschaft und Werkstofftechnik. 2007;**38**(11):934-938

[44] Wagner G, Eifler D. Bewertung der Eignung der Metall-Ultraschallschweißtechnik zum Fügen von Glasfasertextilien und Glasfaserverbundwerkstoffen. In: Hufenbach W, editor. Textile Verbundbauweisen und Fertigungstechnologien für Leichtbaustrukturen des Maschinen- und Fahrzeugbaus. Dresden: SDV-Die Medien AG; 2007. pp. 19-20

[45] Beiß T, Dallner C, Schmachtenberger EM, Wacker M. Vibrationsfügen faserverstärkter Duroplaste. Joinning Plastics. 2007;**1**(2):180-186

[46] Bos M. Deformations- und Dehnungsanalyse von geschweißten Metall/Faser-Kunststoff-Verbunden mit optischen und thermischen Messverfahren. In: Dissertation. Kaiserslautern: Technische Universität Kaiserslautern, Fachbereich Maschinenbau und Verfahrenstechnik; 2009

[47] Velthuis R, Mitschang P, Schlarb AK. Prozessführung zur Herstellung und Eigenschaften von Metall/Faser-Kunststoff-Verbunden. In: 15. Symposium: Verbundwerkstoffe und Werkstoffverbunde. Kassel; 2005

[48] Krüger S, Wagner G, Eifler D. Schweißen artfremder Werkstoffpaarungen - Ultraschallschweißen von metallischen Halteelementen an konsolidierte Glasfaser-Verbundwerkstoffe. Materials Testing. 2004;**46**(3):96-101

[49] Mitschang P, Velthuis R. Process parameters for induction welding of metal/composite joints. In: ECCM-13. Stockholm; 2008

[50] Wilhelm W, inventor; AG B, assignee. Verfahren zur Herstellung von Verbundbauteilen aus Metall und Kunststoff sowie diese Verbundbauteile. Germany patent DE000010133292A1. 2003

[51] Erhard G. Konstruieren mit Kunststoffen. 4th ed. München: Carl Hanser Verlag; 2008

[52] Moritzer E. Spritznieten als neue Organoblech-Metall-Hybridfügetechnik. final report. Paderborn: Industrielle Gemeinschaftsforschung, DVS Forschungsvereinigung, Universität Paderborn; 2021. Report No.: IGF-Nr. 19.796 N

[53] Krassmann D, Moritzer E. Development of a new joining technology for hybrid joints of sheet metal and continuous fiber-reinforced thermoplastics. Welding in the World. 2022;**66**:45-60

[54] Collings TA. Experimentally Determinded strength of mechanically fastened joints. In: Matthews FL, editor. Joining Fibre-Reinforced Plastics. New York: Elsevier Applied Science; 1987. pp. 9-63

[55] Noll TJ. Beitrag zur Entwicklung punktueller Lasteinleitungen und Verbesserung der Versagensanalyse für Faser- Kunststoff-Verbund-Strukturen unter zyklischer Belastung. In: Dissertation. Kaiserslautern: Technische Universität Kaiserslautern, Institut für Verbundwerkstoffe GmbH; 2008

[56] Enßle M. Bearbeitung von Verbundwerkstoffen am Beispiel der Airbus-Passagiertür A350 XWB. Lightweight Design. 2012;**5**:48-52

[57] Kempmann C, Brinkel F, Weinert K. Temperaturbelastung beim Bohren. Kunststoffe International. 2006;(12):72-77

[58] Kempmann C. Fertigung von Innengewinden in Composite-Werkstoffen. In: Weinert K, editor. Spanende Fertigung – Prozesse Innovationen Werkstoffe. Essen: Vulkan-Verlag GmbH; 2005. pp. 372-382

[59] Kroll L. Textilverstärkte Kunststoffbauteile in funktionsintegrierender Leichtbauweise. In: Wintermantel E, Ha SW, editors. Medizintechnik - Life Science Engineering. Berlin, Heidelberg: Springer-Verlag Berlin Heidelberg; 2009. pp. 343-356

[60] Kroll L, Ulke-Winter L. Leichtbau-Hochleistungsstrukturen nach dem Vorbild der Natur. In: Von der Natur lernen: (R)Evolution in der Entwicklung technischer Systeme für den Apparate- und Anlagenbau. Frankfurt am Main; 2010

[61] Abrecht W. Untersuchung der Spannungssteuerung radialer Festigkeitsverteilung in Bäumen. Dissertation. Karlsruhe: Universität

Karlsruhe, Fakultät für Maschinenbau; 1995

[62] Zispe A. Untersuchungen zur lastgesteuerten Festigkeitsverteilung in Bäumen. Dissertation. Karlsruhe: Universität Karlsruhe, Fakultät für Maschinenbau; 1997

[63] Ledermann M. Beiträge zur Optimierung von Faserverbunden nach dem Vorbild der Natur. Dissertation. Karlsruhe: Universität Karlsruhe, Fakultät für Maschinenbau; 2003

[64] Mattheck C, Reuschel D. Design nach der Natur. Physik in unserer Zeit. 1999;**30**(6):253-258

[65] Spickenheuer A, Uhlig K, Gliesche K, Heinrich G, Albers A, Majic N. Steifigkeitsoptimierung von Faserverbundbauteilen für den extremen Leichtbau. In: 12. Chemnitzer Textiltechnik-Tagung. Chemnitz; 2009. pp. 210-219

[66] Lin HJ, Yang SH. Modeling and analysis of composite laminates with continuous fiber around a circular hole. Journal of Composite Materials. 1993;**27**(5):513-525

[67] Gunderson SL, Lute JA. The use of preformed holes for increased strength and damage tolerance of advanced composites. Journal of Reinforced Plastics and Composites. 1993;**12**(5):559-569

[68] Wiedemann J. Leichtbau 2 - Konstruktion. 2nd ed. Berlin, Heidelberg, New York: Springer-Verlag; 1996

[69] Durante M, Langella A. Bearing behavior of drilled and molded-in holes. Applied Composite Materials. 2009;**16**:297-306

[70] Hufenbach W, Adam F, Helms O, Kupfer R. Gestaltung von textilverbundgerechten Fügezonen mit warmgeformten Bolzenlöchern. Kunststofftechnik. 2010;(11):255-269

[71] Ontool GmbH. THERMDRILL - Die kostensparende Alternative zu Schweiß- und Nietmuttern sowie anderen Verbindungstechniken: Produktinformation. 2012. Available from: http://www.ontool.eu/

[72] Shapovalov O, Ost L, Doynov N, Ambrosio M, Seidlitz H, Michailov V. Substitution von metallischen Schubfeldern im Fahrzeugbau durch fügetechnische Integration von FKV-Schalen. final report. Cottbus, Wildau: Industrielle Gemeinschaftsforschung, Brandenburgische Technische Universität Cottbus-Senftenberg, Fraunhofer-Institut für Angewandte Polymerforschung IAP; 2022. Report No.: IGF-Nr. 20649 BR

[73] Ost L, Shapovalov O, Seidlitz H, Michailov V, Doynov N, Kuke F, et al. Simulation of composites' heating. Kunststoffe International. 2023;(5): 60-64

[74] Sec L, R, RS. LVI. On the influence of obstacles arranged in rectangular order upon the properties of a medium. The London, Edinburgh, and Dublin Philosophical Magazine and Journal of Science. 1892;**34**:481-502

[75] Rolfes R, Hammerschmidt U. Transverse thermal conductivity of CFRP laminates: A numerical and experimental validation of approximation formulae. Composites Science and Technology. 1995;**54**:45-54

[76] Bard S, Schönl F, Demleitner M, Altstädt V. Influence of fiber volume content on thermal conductivity in transverse and fiber direction of carbon fiber-reinforced epoxy laminates. Materials. 2019;**12**:1-10

[77] Ost L, Shapovalov O, Seidlitz H, Michailov V, Doynov N, Kuke F, et al. Entwicklung und Analyse einer Fügestrategie für FKV/Metall-Mischverbindungen auf Basis der CMT-Pinschweißtechnik. Joining Plastics. 2023;**17**:28-35

[78] Kuke F, Seidlitz H, Fritzsche S, Kloshek A, inventors; Cottbus-Senftenberg BUoT, assignee. Verfahren zur mechanischen Verbindung eines ersten Werkstücks mit einem zweiten, elektrisch leitfähigen Werkstück. Germany patent DE102019102234A1; 2019

Chapter 4

Ongoing of Energy Saving and Emission Reduction during Fabrication Processing in China's Shipyards

Jiangchao Wang, Bitao Liu, Zhangjing Bao, Wencheng Jiang, Zichao Zhuo, Langxiong Gan and Yaqing Shu

Abstract

Due to the serious situation and deterioration tendency of the atmosphere environment, energy saving and emission reduction are concentrated and desired for each government and international organization. In this research, energy consumption and pollution emission during fabrication processing in shipyards and ocean engineering factories were holistically surveyed for the first time, while the ship industry is the key pillar of manufacturing for national economic development and dominant monitoring objects with severe environment pollution in China. With the visiting and investigation of six representative factories with construction and repair of ship and offshore structure, consumptions of electrical energy as well as chemical energy were summarized for each fabrication processing according with elementary manufacturing flow, which are mainly determined by working load, requirement of quality, and utilization efficiency of energy. Then, various pollutants generated during fabrication procedures were classified and surveyed, while their emission amounts were also summarized by considering their harm level to human health, atmosphere, and ecological environments. In addition, advanced and practical solutions for emission reduction of dust particles and VOCs (volatile organic compounds) were introduced and carried out while the application results were compared with requirements of corresponding laws and regulations.

Keywords: energy consumption, environment pollution, ship industry, VOCs, fabrication processing

1. Introduction

In accordance with the frequent appearance of globally extreme climate and continuous deterioration of human living environment, more and more governments presented much stricter requirement and execution schedule to deal with high energy consumption and massive pollution emission during industrial production, in particular on the manufacturing industry.

In 1992, "United Nations Framework Convention on Climate Change" was internationally established to respond to the harmful influence on economics and human society due to global climate change, which is an elementary framework convention to deal with the global climate problems with international collaboration pattern [1–2]. "Copenhagen Agreement" was later passed in 2009 during the United Nations Climate Change Conference in Copenhagen [3]. Furthermore, in order to promote people to use renewable energy for pollution emission reduction and environment deterioration prevention, most countries reached a global agreement entitled as "The Paris Agreement" to protect Earth's environment in 2015 during the Paris Climate Change Conference, which proposed arrangement and schedule in detail to respond to global climate change [4–5]. However, due to the withdrawal of United States of America and negative implementation of the arrangement without strictly complying with its schedule by some countries, the environmental planning department of the United Nations disappointedly pointed out that pollution emission is much more serious than ever and the target will not be achieved in the "Emissions Gap Report 2019".

With a rapid increase of manufacturing industries' proportion with high-energy consumption in China, the situation of pollution emission and environmental deterioration has become more and more serious in recent years. Thus, national policy of resource saving and environment protection should be strictly fulfilled by means of improved energy utilization efficiency and ecological environment quality [6–12]. Moreover, some laws and regulations for emission pollution reduction and environment protection were published in succession in China [13–16]. Based on the actual situation, local governments, such as in Shanghai city, also promulgated homologous laws and regulations. It can be seen that energy consumption and pollution emission of shipyards has a serious influence on ecological environment protection and living environment improvement, which will also postpone the implementation of policy regulations with low carbon and environmental protection. Furthermore, it will even impede the fulfillment of international commitments, prestige and reputation of responsible great power.

Shipyard and ocean engineering company, as typical manufacturing industry with high pollution emission have become key monitoring objects by local environment-protecting departments. Most investigation is concentrated on the shipping process, which is a popular commercial activity in modern economics and generally causes remarkable air pollution [17, 18]. Meanwhile, shipbuilding and offshore construction industries, as a typical representative of entity economy and traditional manufacturing, are a key pillar industry of national economics and defense security in China, which also play an essential role in the promotion of economic development and improvement of people living. However, shipbuilding and offshore construction industries in China almost adopt the conventional extensive production model for fabrication processing at present with typical characteristics such as labor-intensive and poor working conditions, heavy work load, and roughly less profits, which is already and totally not suitable for the requirement of high-quality development of advanced ship and offshore manufacturing [19]. Moreover, this kind of situation resulted in large requirement of energy consumption, lower utilization efficiency of energy, and massive emission of various wastes and pollutions during individual fabrication phase, which causes serious harm on living environment and human health. Celebi and Vardar [20] examined the VOCs emission from indoor and outdoor painting processes for the shipbuilding and ship repair industry. In addition, the total area of painted surface and total paint consumption of 3500 DWT oil/chemical tanker and a general bulk carrier were calculated and compared. Chung and Lee [21] measured ambient concentrations of hazardous air pollutants such as VOCs and aldehydes in residential areas nearby small- and medium-scale shipbuilding companies, while distribution characteristic and influential factors of VOCs emission were also presented. Malherbe and Mandin [22] concentrated on VOCs due to painting of ships' external surfaces in building or repair shipyards, and a health-risk assessment based on measured data of Chantiers de l'Atlantique was carried out to evaluate the risks by inhalation for people living near the site.

According to corresponding laws and regulations, national and local environment-protecting departments have already investigated and administratively punished the shipyard and ocean engineering company with excess pollution emission or environmental pollution behavior. Due to the lack of a practical solution, treatment results of energy saving and emission reduction are still not remarkable. In order to survey the current situation of pollution emission on shipbuilding and ocean engineering company in China, 18 punishment cases in recent years had been summarized beforehand as shown in **Figure 1** based on public information searching of national environment-protecting department, which are almost due to excess emission of different pollutants such as volatile organic compounds (VOCs), solid wastes, and industrial sewage or lack of environmental assessment.

During the various fabrication procedures, the fabricated object, manufacturing method, as well as employed equipment are all different; rated power and energy conversion efficiency for individual equipment are also different. Meanwhile, pollution emission during shipbuilding is also complicated due to different pollutant types, harm levels, and its treatment solutions. Therefore, the energy consumption and pollution emission are both difficult to summarize and compare due to different types and amounts.

In this research, fabrication procedures such as plate cutting, flame heating, welding, and painting during shipbuilding and offshore construction were sequentially introduced with consideration to the types of energy consumption and pollution emission. Rated powers of employed equipment as well as energy consumption according to the shipbuilding procedure flow were surveyed and examined from domestic shipyards and ocean engineering companies, while different fabrication methods for identical manufacturing purpose were also compared from energy consumption, fabrication accuracy and cost, as well as limitation of each fabrication methods. In addition, currently dominant procedures with higher energy consumption or lower energy utilization were then pointed out, and the factors with significant influence on energy consumption were also examined. Practical approaches based on advanced fabrication method and computer-aided manufacturing with computational processing mechanics were introduced and employed for energy saving. According to the pollutant type, the current situation about pollution emission during different fabrication procedures was then summarized with detailed survey. In particular, VOCs and smoke-dust particle were concentrated due to their massive emission and serious harm on human health, while not only their emission characteristic and influential factors but also the treatment solution such as advanced technology or additional processing were presented.

Above surveyed data, current characteristic, and their influential factors can support a reliable solution of energy saving and emission reduction with basic data and reference. In addition, based on the concept of energy saving, environment protection as well as intelligent manufacturing, technology updating of fabrication processing and scientific management on shipbuilding phase with features such as higher energy consumption, lower energy utilization, massive amount of pollution emission, severe environment pollution and serious harm on human health, will be then achieved.

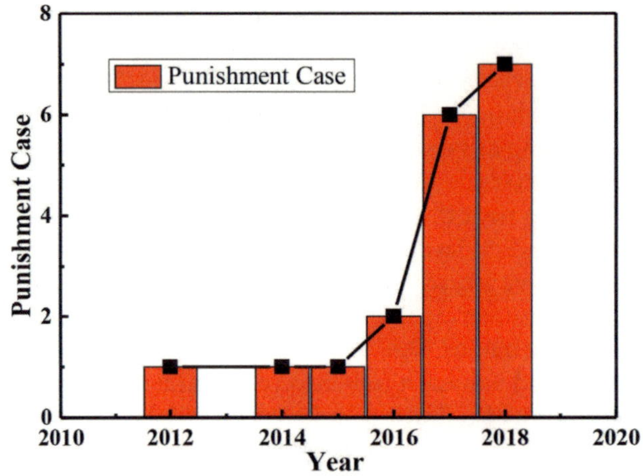

(a) Punishment case with different year

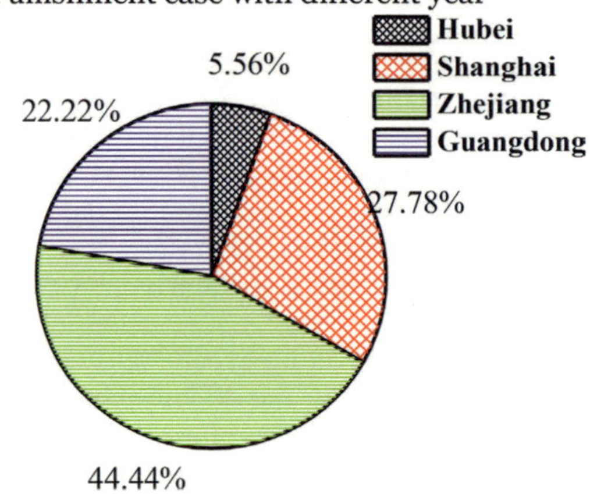

(b) Punishment case with different Province

Figure 1.
Eighteen punishment cases of ship industries in China. (a) Punishment cases in different years. (b) Punishment cases in different provinces.

Moreover, the establishment of HSE (health, safety and environment) management system considering health, safety, and environment together was also practiced, whose target is to guarantee human health and safety with less environment pollution during each shipbuilding phase. Eventually, advanced fabrication processing and operation standard with optimization of shipbuilding processing will be proposed to enhance the level of fabrication-processing technology and detailed management, while energy utilization efficiency and productivity, high-quality development capacity of shipbuilding industries, and competition of international market will also be enhanced.

2. Fabrication procedures during shipbuilding

With the necessary requirement of launching processing for ship moving from construction position to operation region, shipyards are almost constructed at coastal areas nearby an inland river and ocean. By internet search and actual visiting survey, domestic shipyards generally locate at shore of Yangtze River in Hubei province, Jiangsu province and Shanghai city, as well as shore of Pearl River in Guangdong province; coastal area in Liaoning province, Tianjin city, Shandong province, Zhejiang province, and Fujian province also have lots of shipyard and ocean engineering companies. Abovementioned regions with the shipbuilding industry are almost areas with developed and prosperous economics in China.

In addition, energy consumption will be influenced not only by fabrication processing and its energy utilization efficiency but also by other factors such as location and condition of shipyard, regional policy, construction capacity, as well as the type of manufacturing products such as bulk carrier, container ship, and other high-value and high-tech added vessels. Therefore, in order to obtain much more representational data of energy consumption and pollution emission of China shipyards, there are six shipbuilding and ocean engineering companies with visiting and investigation as shown in **Figure 2**.

With the viewpoint of processing flow in modern shipbuilding mode, shipbuilding usually includes three fabrication phases: hull construction, outfitting, and painting [19]. In addition, shipbuilding and offshore construction almost have the identical fabrication procedure flow as illustrated in **Figure 3**. Typical fabrication processing for hull construction are steel pretreatment, plate edge processing, plate forming processing, assembling-welding processing, as well as undocking and launching. Ship repair process has a little bit different feature comparing with structure construction such as outside surface clearance due to plankton attachment or marine corrosion, as well as arbitrary emission of VOCs and dust particles, repair of damaged structure with cutting and welding, and replacement and debugging of damaged equipment. Meanwhile, there are some problems for the domestic shipbuilding industry compared with other advanced manufacturing, which has a longer fabrication schedule, higher fabrication cost, higher fabrication accuracy, and poorer working environment.

Towards to survey and investigation of energy consumption and pollution emission, hull construction and painting were dominantly examined. In particular,

Figure 2.
Location of investigated shipyard and ocean engineering companies.

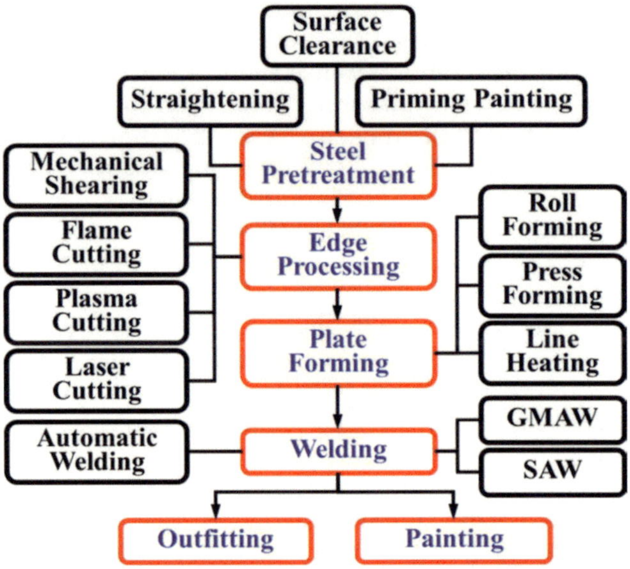

Figure 3.
General fabrication procedure flow in ship industries.

painting processing as a surface engineering is carried out to prevent corrosion and plankton attachment with purpose of protecting the ship structure during service period. As summarized in **Table 1**, energy type and possible pollution emission during individual fabrication processing for ship construction and repair were listed. It can be seen that various energy will be employed for different fabrication phase or manufacturing method, while electrical energy and chemical energy are the dominant energy form. Meanwhile, lots of pollutant such as dust particles, noise, VOCs, and solid waste will be generated and emitted into natural environment during the individual shipbuilding phase as summarized in **Table 1**, which will have devastating influence on the atmosphere and ecological environment. In actuality, most fabrication procedures can be carried out in a workshop with good condition while pollutants such as VOCs can be gathered and recycled for environment protection with lower cost. However, some fabrication procedure will be done outside such as at a dock or wharf, and pollution emission will become much more serious. Thus, much more attention is required on controlling pollution emission; advanced technology should be proposed and employed for reducing pollution emission with lower cost and higher efficiency.

3. Surveys of energy consumption and saving

As shown in **Table 1**, various energy will be employed for different fabrication processing due to working objective, working purpose, processing approach, and applied equipment, while working hours will be determined by fabrication loading for evaluation of energy consumption.

In general, the dominant energy is electrical energy, while fabrication processing for manufacturing demand of steel plate and ship structure are achieved by different type of electrical-mechanical equipment with electrical energy supply.

Processing	Fabrication approach	Energy type	Pollution emission
Steel pre-treatment	Mechanical straightening	Electrical Energy	None
	Surface clearance of shot blast	Electrical energy	Dust particle, noise
	Priming painting protection	Electrical energy	VOCs
Plate edge fabrication	Mechanical shearing	Electrical Energy	None
	Flame cutting	Chemical energy	Dust particle Solid waste
	Physical cutting	Electrical energy	Noise, solid waste
Plate forming	Roll forming	Electrical energy	None
	Press forming	Electrical energy	None
	Thermal forming	Chemical energy	Effluent liquor
Welding and burnishing	GMAW with CO_2	Electrical energy	Dust particle, arc light, solid waste
	SAW	Electrical energy	None
	Burnishing	Electrical energy	Dust particle, noise
Ship painting	Spray coating	Electrical energy	VOCs, noise

Table 1.
Energy type and pollution emission during ship construction and repair.

Meanwhile, fabrication equipment due to different rating power and conversion efficiency of electrical energy will result in significant difference of energy consumption for each fabrication procedure. In addition, consumption of electrical energy during shipbuilding can be evaluated with rating power of fabrication equipment and working hours, which unit generally is KW*h. Furthermore, chemical energy is also employed for fabrication processing of steel plate and ship structure. Taking plate cutting and bending with flame heating as example, flame gas such as acetylene, propane, and methane (natural gas) are burned to heat a steel plate until its combustion temperature, and these flame gases are also burned to generate heat, temperature gradient in plate thickness direction, as well as bending moment for plate bending. During plate flame cutting, oxygen is another supplied gas for steel burning as violent oxidation reaction after reaching combustion temperature with plate preheating.

In addition, massive gases such as carbon dioxide and argon will be consumed during the processing of gas shielded arc welding while these gases can protect welding arc with stable droplet transition, favorable welding process, and excellent welding quality. In actuality, consumption of chemical energy as well as gas is difficult to be measured and evaluated due to local adjustment with manufacturing situation during previously mentioned fabrication processing.

3.1 Steel pretreatment processing

Steel pretreatment processing as the first shipbuilding phase has three fabrication contents such as plate straightening, surface rust cleaning, and priming paint protection. **Figure 4** shows the typical equipment for steel pretreatment processing, which

Figure 4.
Typical equipment for steel pretreatment.

are usually utilized together as a continuous line of mechanization production in a shipyard and ocean engineering company.

Energy consumption of steel pretreatment line is electrical energy, and their rating powers are generally about 1000KW for investigated shipbuilding companies. In addition, fabrication processing such as surface rust cleaning and priming paint protection will consume massive electrical energy due to rating power of equipment and fabrication loading, which are usually considered as processing phase with high energy consumption during shipbuilding.

3.2 Plate edge processing

After steel pretreatment processing, plate edge processing, including plate cutting and welding groove working, is then carried out according to the design and arrangement based on the geometrical feature of various pieces. With different physical mechanisms, there are generally three approaches for plate cutting during shipbuilding, which are mechanical cutting, chemical cutting with flame heating, and plasma cutting.

As shown in **Figure 5**, this is typical equipment for plate edge processing, while flame heating as the most popular approach is employed for plate cutting and welding groove working. **Figure 5** also shows the typical equipment for plate cutting with plasma.

Figure 6 compares the rating power of different cutting methods during shipbuilding. It can be seen that different cutting approaches will be employed for plate cutting due to plate thickness, and the rating power of flame cutting is much less than

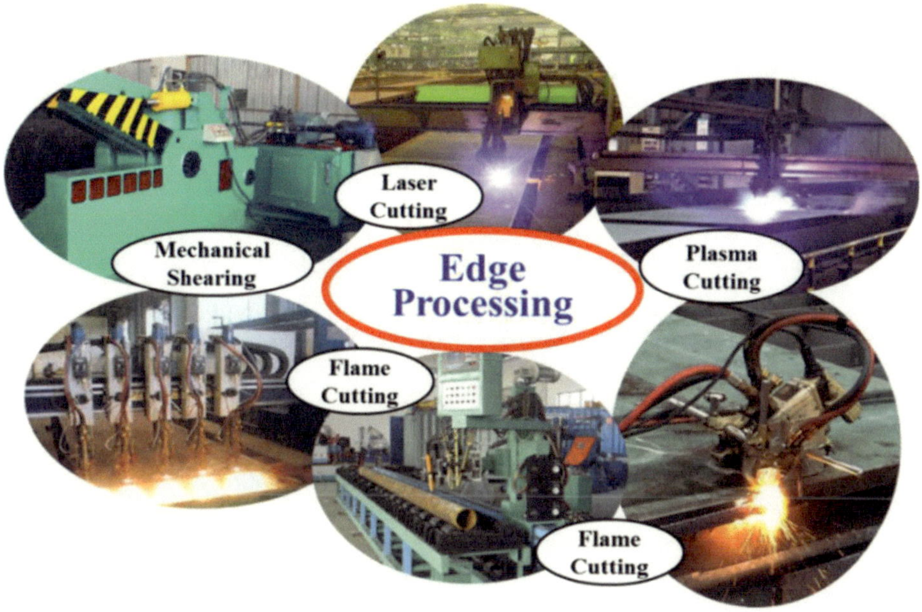

Figure 5.
Typical equipment for plate edge processing.

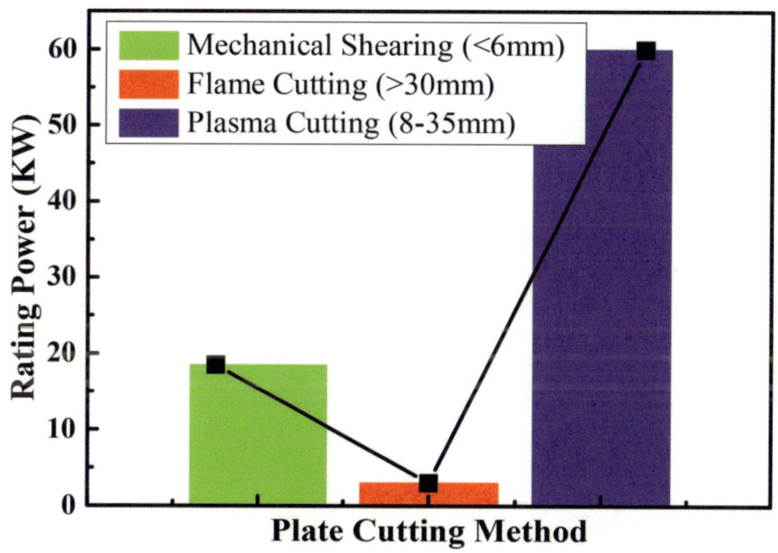

Figure 6.
Comparison of the rating power of typical cutting equipment.

that of plasma cutting, which is the result of massive consumption chemical energy of combustible gases. Moreover, the material processing workshop of CIMC Raffles offshore engineering company was visited and investigated, while cutting capacity is demonstrated to be about 2,412m for each day as shown in **Figure 7**.

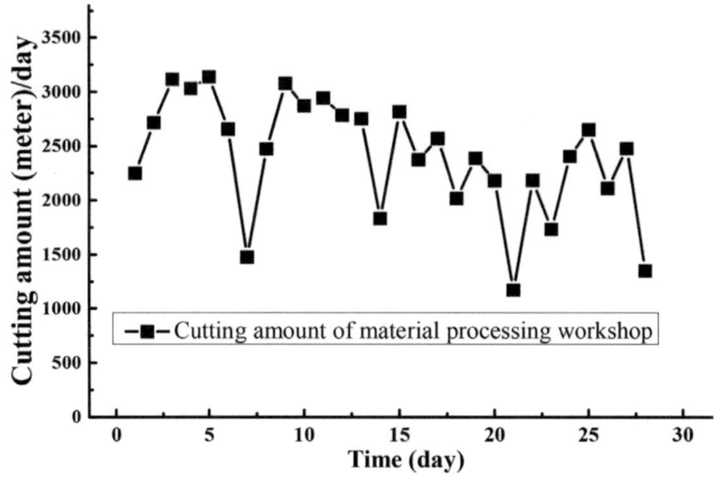

Figure 7.
Cutting capacity of material processing workshop (CIMC Raffles offshore engineering company).

3.3 Plate bending processing

In order to guarantee excellent hydrodynamic performance of advanced ship, ship hull structure is designed with various curvatures, which can be obtained with plate forming processing in shipbuilding [23]. Currently, plate forming for curved hull structure can be achieved by either cold forming with rolling and mechanical pressing or thermal forming such as oxygen-acetylene flame heating, laser heating, and induction heating as shown in **Figure 8**. During roll bending, displacement of upper roller with moving down and distance between the two lower rollers will determine the bending curvature, while spring-back magnitude should be considered due to elastic-plastic mechanical behavior. In general, plate forming with single curvature is achieved by roll forming with three-roller, and press forming is used for plate forming with double curvatures as well as complex curved geometry.

As mentioned above, energy consumption of cold forming processing can be evaluated with the rating power of employed equipment and working hour. **Figure 9** summarizes the rating power of equipment with roll-forming and press-forming approaches. In addition, energy consumption of cold-forming processing for ship hull construction is generally not too much due to expensive equipment and weak flexibility.

Besides plate cold-forming processing with rolling and pressure, plate forming also can be carried out with thermal processing. In addition, flame heating by burning of acetylene, propane, and methane (natural gas) or induction heating can be employed to heat plate surface. Then, the temperature gradient and nonuniform distribution of residual plastic strain in plate thickness direction can be generated, while bending moment is also generated to achieve out-of-plane bending deformation. In **Figure 8**, plate bending is carried out with a skilled worker by means of flame heating, while water cooling is usually employed to enhance the temperature gradient and bending moment, as well as bending processing efficiency.

Comparing with the limitation of plate cold forming, plate thermal forming can be employed to obtain curved plate with arbitrary geometry such as variable curvatures, complex double curvatures. Therefore, geometrical shape and fabrication

Figure 8.
Typical fabrication processing for plate bending.

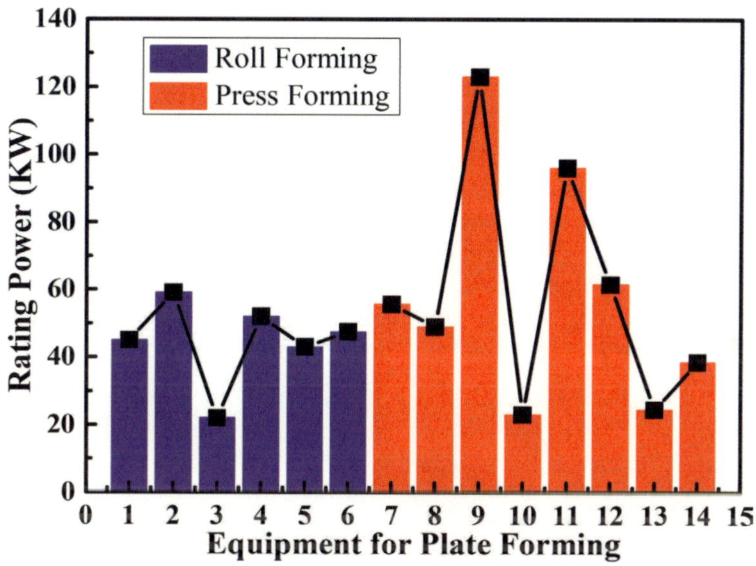

Figure 9.
Equipment and their energy consumption for plate cold bending processing.

accuracy after plate bending should be measured and checked with various templates according to objective curvature; also, non-touch laser scanning technology is already employed with high efficiency and precision for cured plate measurement in some advanced shipyards.

3.4 Assembling and welding processing

Welding with assembling as the most essential fabrication processing in shipbuilding is employed to gradually fabricate component, structural piece, ship section, and block with joining approach. By definition, welding means a joining activity with molecular bonding together for similar or dissimilar metal parts, while heat or pressure is applied, and sometimes both applications are carried out [24]. Working hours during welding processing will take about 50–60% of the shipbuilding schedule due to massive production loading and relatively lower productivity with less automatic equipment such as welding robot. Nowadays, employed welding approach in Chinese shipyard is manual welding, GMAW (gas metal arc welding) with CO_2, and SAW (submerged arc welding).

As shown in **Figure 10**, GMAW with CO_2 can be generally observed for stiffened ship panel welding in assembling fabrication workshop. Also, a welder should wear working protective clothing and shielded face mask to avoid radiation of welding arc on skin cauma. Meanwhile, SAW is usually employed for thick plate butt-welding by semiautomatic equipment as shown in **Figure 10**. SAW can support enough fabrication requirements such as quality of welding seam and mechanical performance, while welding equipment with large volume and high cost is usually employed for ship hull structure fabrication with thick plates.

It can be seen that working condition of gas shielded arc welding with CO_2 is roughly terrible, and massive smoke particles are generated, which have serious pollution influence on atmosphere environment. However, gas-shielded arc welding has better flexibility and lower cost compared with SAW to implement complex welding requirement during the fabrication of various ship structures, while SAW is carried out with certain limitation for wide application due to the requirement of moving car and its guide rail facilities.

In the view of energy consumption and welding heat input, SAW has large linear energy and can be employed for thick plate butt-welding with large heat input for high productivity. Heat efficiency of SAW is almost over 90% due to the protection action of weld flux, while electrical energy can be well converted to heat energy for metal fusion welding. This weld flux, which guarantees stable processing of arc welding, can be recycled to reduce fabrication cost. Meanwhile, heat efficiency of gas shielded arc welding with CO_2 is about 60–80%, while partial electrical energy is converted to arc light and heat energy with loss to ambient air as compared in **Figure 11**.

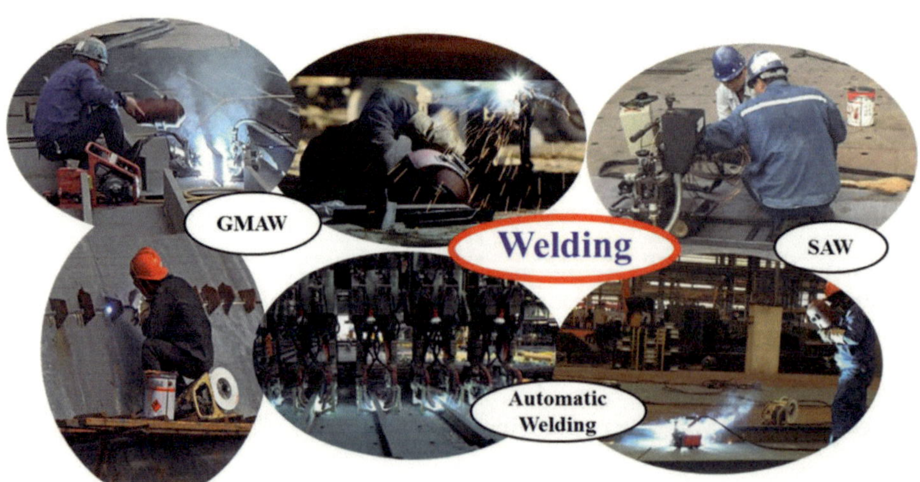

Figure 10.
Typical welding method during shipbuilding.

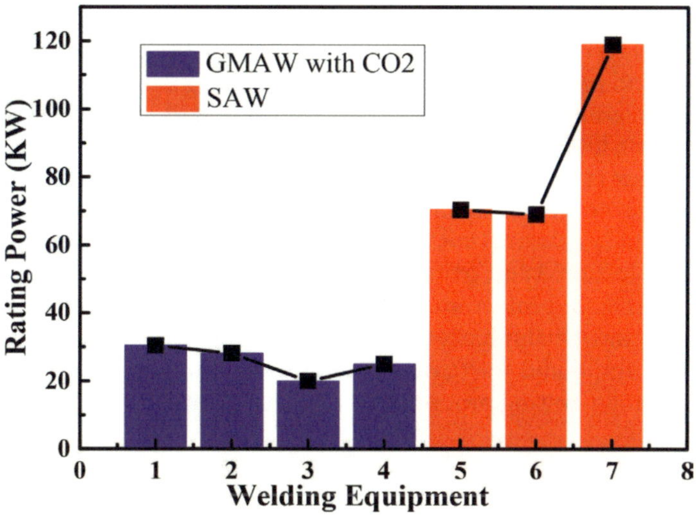

Figure 11.
Comparison of rating power of GMAW and SAW.

3.5 Plate and structure painting processing

Besides priming painting protection during steel pretreatment, painting processing during ship section fabrication in workshop and ship block joining in outside dock is sequentially carried out with anticorrosion and antifouling processing on a steel plate surface to improve the service life of a ship and offshore structure. Painting processing is achieved by supporting equipment to uniformly spray coating material with protected purpose on the inside and outside surfaces of ship structure.

In general, paining processing is always carried out in a closed workshop for atmosphere-pollution prevention due to massive emission of hazardous waste gas. Rating power of painting equipment is about 1000 kW with investigation, while energy consumption of painting equipment is roughly large comparing with that during previous fabrication processing. Also, large fabrication loading about painting is desired due to different protection requirement of ship region under various external environment loadings.

4. Survey situation of pollution emission

As mentioned above, there are various fabrication procedures such as steel plate cutting, bending, and welding in shipyard; in addition, pollutants such as dust particles (PM2.5, PM10), VOCs, noise, sewage, as well as other solid and liquid waste will be also generated and emitted into nature, which has a serious influence on the atmosphere and ecological environment. In particular, VOCs as well as hazardous wastes such as waste paint barrel and waste paint residue will be inevitably generated during painting procedure. Steel plate cutting and welding will dominantly generate smoke and dust particles, and each fabrication procedure inevitably generates noise pollution.

4.1 Solid and liquid waste

Leftover bits and pieces of ship steel generated during each shipbuilding phase can be recycled as temporary stiffener for reinforcement of structural stiffness. In general, there are lots of special collection boxes for gathering leftover bits and pieces in the fabrication workshop. Meanwhile, according to treatment regulation of hazardous wastes in shipbuilding industry, some solid and liquid waste with pollution effect on natural and ecological environment generated during fabrication procedure should be strictly gathered and treated by the company with professional certification with environmental protection. As compared in **Figure 12**, emission amount of pollutants such as waste paint barrel, paint waste as well as waste paint residue, waste activated carbon, and liquid waste were listed for surveyed domestic shipyard during the construction and painting procedure of ship structure.

4.2 Dust particles emission (PM2.5/PM10)

Emission of smoke-dust particles during ship and offshore construction is a common environment-pollution behavior, which will not only influence the atmosphere environment but also hurt the health of a worker. In general, massive smoke-dust particles will be generated throughout the entire phases of surface cleaning of steel plate, plate cutting, as well as welding.

Plasma cutting with lots of advantages is employed as popular cutting approach for plate edge fabrication in shipyard; however, massive smoke-dust particles will be also generated during plate cutting as shown in **Figure 13**, which causes serious pollution in the atmosphere environment. Meanwhile, welding process is another fabrication phase with the generation of massive smoke-dust particles as shown in **Figure 14**, and welding procedure as essential fabrication phase during shipbuilding roughly consumes about 50–60% working time of the total production

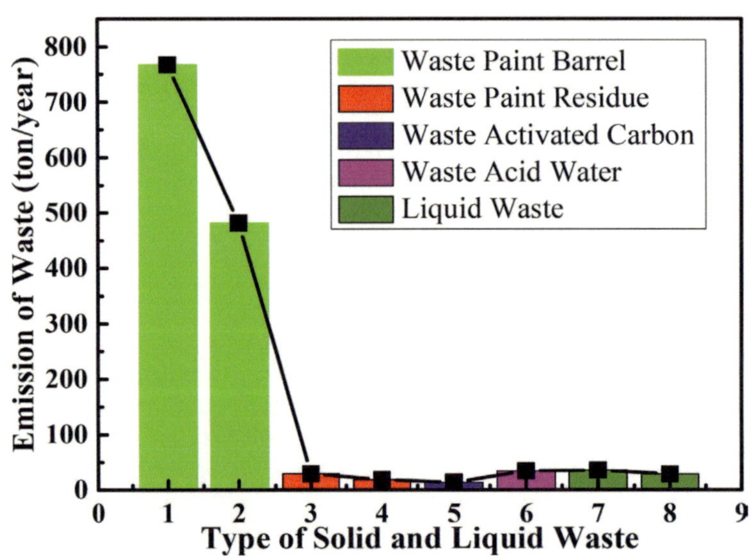

Figure 12.
Emission amount of solid and liquid waste during shipbuilding.

Figure 13.
Dust particles emission during plasma cutting.

Figure 14.
Dust particles emission during welding.

schedule. For ship and offshore structure welding, manual welding and SMAW with CO_2 are always employed due to their advantages such as lower cost and flexible operation feature.

Also, welding smoke-dust particles are difficult to gather and recycle because of various welding situation, which will have serious effect on atmosphere pollution and human health. SAW with large welding heat input is usually employed for thick plate butt-welding without emission of welding smoke-dust particles, while mobile equipment are required to limit their wide engineering application. In particular, the outside surface of a ship structure is required to be cleaned and polished during ship repair, while abrasive blasting at wharf or dock will generate massive dust particles and serious pollution of the atmosphere and environment due to arbitrary emission.

4.3 Volatile organic compounds

VOCs mean any compounds of carbon with volatility feature, which have an initial boiling point less than or equal to 250°C measured at a standard atmospheric pressure of 101.3 kPa [25–27]. VOCs will be generated and emitted during the painting processing for shipbuilding, which has the hazardous feature of strong volatility, pungent odor, and high poisonousness. In addition, VOCs have a harmful influence on human health, in particular serious harm on respiratory system of human, which even imperils the life safety of human. Thus, emission reduction of VOCs has significant influence and meaning for both the environment and human life. VOCs are already considered as important controlling and monitoring objective among various emitted contaminants during shipbuilding.

In the phase of steel pretreatment, priming paint protection will be carried out after surface rust cleaning with generation of VOCs and atmosphere and environment pollution. During the painting process of ship and offshore construction, massive paint will be consumed, and lots of VOCs will then be generated with serious pollution of the atmosphere and environment. In general, about 70–75% of the total paint will be consumed during the preprinting of ship structure in the fabrication workshop, and the remainder 25–30% of the total paint will be consumed during wharf painting. As shown in **Figure 15**, consumption amount of various paints and emission amount of VOCs for several domestic shipyards are listed, while their consumption and emission are mainly determined by total working load of shipbuilding. In addition, average density of VOCs is assumed as 0.8 g/l, and VOCs content of various paints can be evaluated as 650 g/l, which results from the VOCs content standard

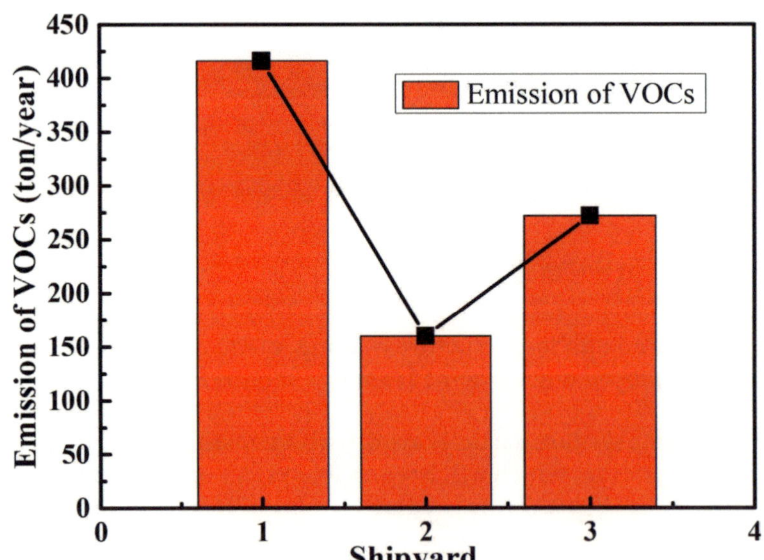

Figure 15.
Emission of VOCs in domestic shipyards.

(unit: g/l) of paint in use for shipbuilding [28]. Moreover, VOCs content of paint in use for shipbuilding can be calculated with Eq. (1):

$$C_{\text{VOCs}} = \frac{C_P V_P + C_t V_t}{V_P + V_t} \tag{1}$$

Where CVOCs (g/l) mean the VOCs' content of paint in use for shipbuilding; C_p and C_t (g/l) mean the VOCs' content of paint for shipbuilding and diluent, respectively; and V_p and V_t (l) mean the used amount of paint for shipbuilding and diluent, respectively.

4.4 Noise pollution

During each fabrication process of shipbuilding, noise is inevitably generated, while its pollution occurs more or less [29]. With the manual measurement with decibel meter, there is serious noise pollution in shipyards. Moreover, the noise at operation position of worker was measured during different fabrication processing as shown in **Figure 16**. Noise pollution will have harmful influence not only on the listening capacity and psychological health of a worker but also on the ordinary daily life of people living in staff quarters and nearby residential quarters. Therefore, necessary safeguard approach should be considered and employed such as protective clothing for workers and sound insulation equipment at the boundary of a factory.

5. Progresses and advanced solution for emission reduction

In order to achieve collaboration of high-quality development of shipbuilding industries and environment protection, actual fabrication processing for ship and

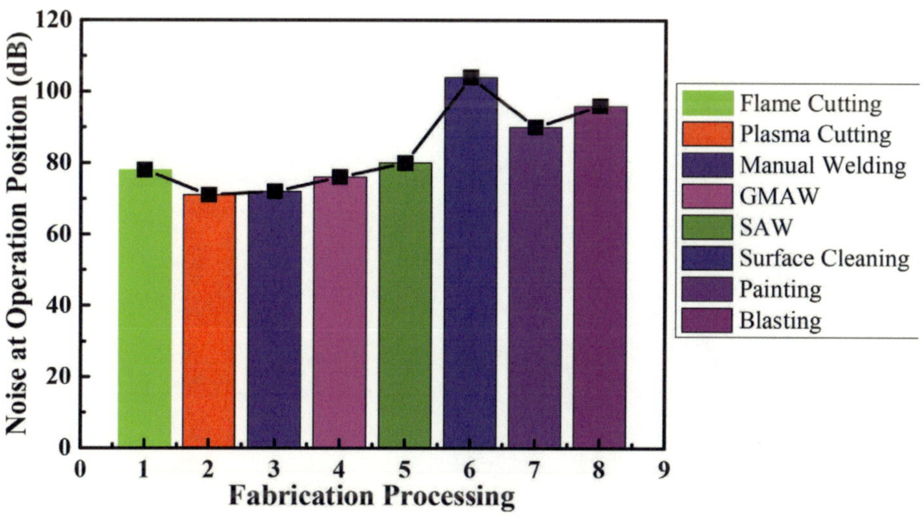

Figure 16.
Measured noise data for the ship industry.

offshore construction should be carried out according to corresponding policy and regulations.

At the beginning, most shipyards pay more attention to environmental governing and promotion of pollution-emission reduction in the factory district and fabrication workshop, which will improve and modify the working environment, as well as obtain certification of national environment protecting department. Various banners for environmental protection and pollution-emission reduction are hung in a conspicuous position of a factory district and fabrication workshop, which can constantly remind all workers to pay attention on environmental protection during fabrication processing. Moreover, these banners will increasingly enhance the consciousness and responsibility of environmental protection for each manufacturing staff member.

5.1 Solution for emission reduction of dust particles

Among the fabrication procedures during ship and offshore construction, the dust particles will be generated and emitted to the atmosphere and environment in the phases of surface cleaning of steel plate, plate cutting, as well as welding.

In order to reduce the emission of dust particles for environment protection, advanced cutting processing such as underwater plasma cutting and abrasive water jet (AWJ) cutting with high pressure were applied, while thermal deformation can be effectively reduced for accurate cutting. Ship plate steel is cut by AWJ with high pressure, which is an advanced green fabrication approach with development of water jet technique and high-speed grinding technique [30]. Meanwhile, reduction of smoke-dust particles emission can be achieved by modifying and upgrading flame-cutting equipment as well as plasma cutting equipment during the plate-cutting procedure, while recycling device of smoke-dust particles is usually installed. In addition, exhaust system with high power, as shown in **Figure 17**, is installed under the cutting platform to totally gather and recovery smoke-dust particles while avoiding arbitrary emission and atmosphere pollution. As usual, dust cover around cutting nozzle is employed to enhance the recycling efficiency of smoke-dust particles during cutting procedure.

Figure 17.
Exhaust system with high power for recycling smoke-dust particles during the cutting procedure.

For reduction of welding smoke-dust particles, dust catcher with movable device as shown in **Figure 18** is employed in a shipyard. For the center welding position, welding smoke-dust particles can be gathered and recycled with movable dust catcher, while atmosphere environment can be well protected by means of avoiding arbitrary emission. However, the operation inconvenience and huge requirement are disadvantages that limit their wide engineering application only for some positions with good welding configuration. An exhaust fan, as shown in **Figure 18**, is also installed in the more position of fabrication workshop to effectively gather and recovery smoke-dust particles for arbitrary emission elimination during plate cutting and structure welding. Moreover, welding wire with advanced chemical components and high cost is another option for less emission of welding smoke-dust particles and environment protection in some shipyard and ocean engineering companies.

With the above solution for reduction of smoke-dust particles emission during plate cutting and structure welding processes, the vast majority of smoke-dust particles can be gathered and recycled, and the amount of smoke-dust particles emission can basically satisfy the emission standard of laws and regulations [31].

5.2 Solution for emission reduction of VOCs

In order to satisfy the requirement of atmosphere and environment protection, reduction of pollution emission is usually considered during the ship and offshore construction. In particular, VOCs during the priming painting protection of pretreatment and painting will be generated and emitted to atmosphere, which has serious harm to human health and is considered as the essential pollutant of the environment.

Currently, there are some practical solutions for emission reduction of VOCs and enhancement of atmosphere environment quality, which will be employed due to

Figure 18.
Movable dust catcher for welding process and exhaust fan in workshop.

configuration of organic exhaust gas such as component percentage, VOCs concentration, as well as its amount and flow velocity. During painting processing for ship and offshore construction, treatment technology of organic exhaust gas including VOCs is generally employed to control and reduce emission amount with terminal governance, which are basically classified into two classifications: recycling technology and removing technology with combustion decomposition of thermal oxidation or biochemical reaction of VOCs [26, 27]. Nowadays, adsorption recycling technology, thermal oxidization technology with catalytic combustion, and biological technology are the main top three treatment technology employed for VOCs emission treatment in shipyard, whose percentages of market share are 38%, 22%, and 15%, respectively. Meanwhile, other advanced technologies such as biological technology, photochemical catalysis technology, membrane technology and plasma destruction technology, and photochemical catalysis technology are gradually employed to effectively reduce the VOCs amount of organic exhaust gas and to rapidly reach emission standard.

Based on the applied technology for VOCs treatment, recycling technology with pressure swing adsorption, thermal swing adsorption, adsorption with activated carbon, and condensation is usually employed to gather VOCs for exhaust gas with high concentration of VOCs, while the remaining amount of VOCs should be reduced to the emission limit of ship industry [28]. For the exhaust gas with medium concentration of VOCs, recycling technology can be employed; meanwhile, thermal oxidization technologies due to combustion with regenerative thermal oxidizer (RTO) and regenerative catalytic oxidation (RCO) are also employed to clear organic exhaust gas, while VOCs will be converted to CO_2 and H_2O without harmful effect to human, and these gases will be then emitted into the atmosphere. Heat generated during VOCs combustion should be recycled for preheating the organic exhaust gas to be treated with thermal oxidization technology. For the exhaust gas with low concentration of VOCs, thermal oxidization technology due to combustion is usually employed, while pressure swing adsorption and condensation are always carried out in advance to obtain VOCs with high concentration for effective combustion.

For the shipyard and ocean engineering construction industry, emission of VOCs during painting processing has some characteristics such as diversity of generation sources, complicated components, lower concentration, valueless or higher cost of recycling, as well as massive amount with distribution. Therefore, adsorption technology based on activated carbon with honeycomb form with large specific surface area is usually employed for recycling of VOCs during treatment of organic exhaust gas with low concentration, and exhaust gas by means of adsorption clearance with less VOCs will be emitted with external vent pipeline as shown in **Figure 19**. In the painting workshop, advanced treatment approach with adsorption-condensation-desorption-catalytic combustion is generally carried out for thoroughgoing VOCs clearance, while heat generated during VOCs combustion is gathered to preheat the subsequent organic exhaust gas for cost reduction without additional consumption of fuel. As shown in **Figure 20**, processing line and its corresponding equipment are demonstrated in shipyard. In detail, VOCs with low concentration is condensed with activated carbon with honeycomb form beforehand, and then, desorption processing is carried out to obtain condensed VOCs with higher concentration, which will be burned by means of catalytic combustion. In addition, noble metal (such as platinum and palladium) as catalysator is applied to burn VOCs at lower temperature, and CO_2 and H_2O without harm to human health will be generated and emitted into atmosphere.

Figure 19.
Adsorption and gathering equipment to deal with organic exhaust gas with low concentration VOCs.

The detailed information and technological parameters of equipment for clearance of low concentration VOCs in painting workshop are listed in **Table 2** while activated carbon with honeycomb form, noble metal (such as platinum and palladium) as catalysator, as well as technological parameters of absorption device and catalytic combustion device are also introduced.

Among all thermal oxidization technologies, RTO and RCO are the most common technology for VOCs clearance due to steady application and best results.

In detail, RTO technology can heat the organic exhaust gas with VOCs to the temperature of above 760°C and make VOCs decompose with Oxidation combustion to be CO_2 and H_2O while the generated heat during VOCs combustion will be stored in a special heat accumulator. Then, this regenerative heat will be employed for preheating the subsequent organic exhaust gas with thermal oxidization combustion, which requires the processing working of oxidation combustion to continuously maintain as soon as possible. In addition, fuel consumption can then be reduced for heating of organic exhaust gas, while fuel consumption is determined by flow velocity of organic exhaust gas, concentration of VOCs, thermal storage capacity of heat accumulator, as well as surface heat loss due to convection and radiation. It is suitable to deal with organic exhaust gas with VOCs of 100–3500 mg/m^3, and decomposition efficiency of VOCs is about 95–99% with convenient operation and long service life; however, the weight of equipment is huge due to the application of ceramic heat accumulator, which is usually located in the outdoor due to its large volume. Meanwhile, investment of RTO is much more expensive

With catalysator of noble metal (Pt/Pd), RCO technology can effectively decompose the VOCs by means of oxidation combustion into CO_2 and H_2O in lower temperature of 200–400°C for solving pollution of organic exhaust gas. RCO has advantages such as lower requirement of fuel consumption and high quality of VOCs clearance with 97%, and it is suitable to deal with organic exhaust gas with flow velocity of 5000–200,000 m^3/h by means of convenient operation. In addition, a special catalysator of noble metal is almost impregnated into storage ceramic material with honeycomb form for both lower fuel consumption and less cost of equipment, while the best clearance effect can be achieved and efficiency of thermal recovery is higher than 95%. Moreover, both thermal oxidizer (RTO) and RCO require expensive cost for corresponding equipment installation, and massive electrical energy will

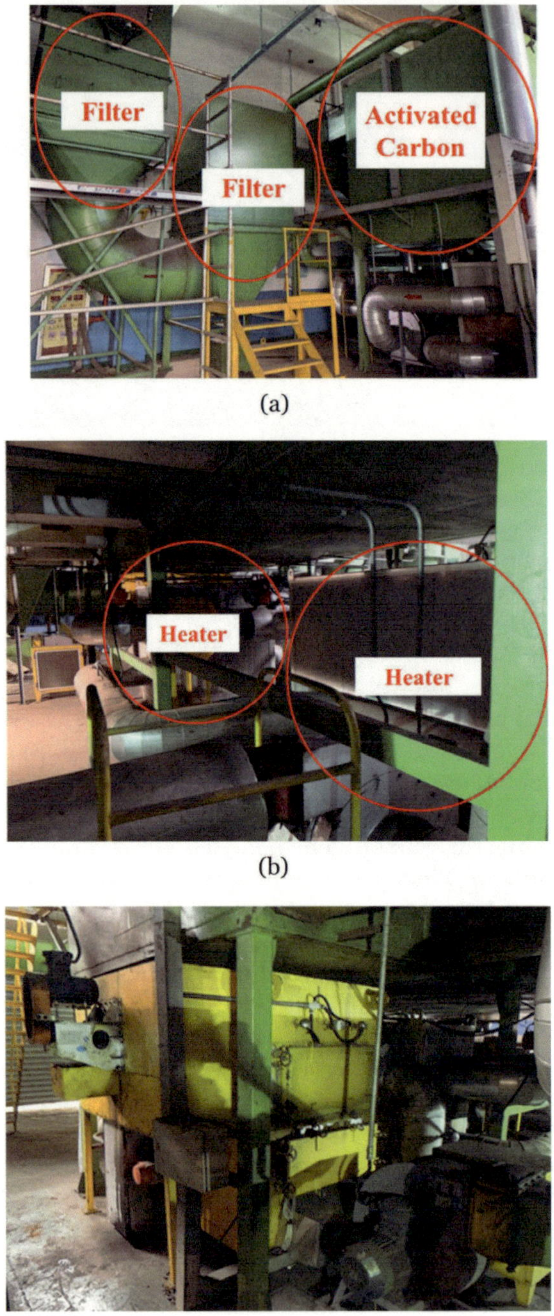

(a)

(b)

(c)

Figure 20.
VOCs clearance with adsorption–condensation–desorption–catalytic combustion processing and corresponding equipment. (a) Adsorption–condensation device based on activated carbon with honeycomb form on organic exhaust gas with low concentration VOCs. (b) Desorption of condensed VOCs form activated carbon with honeycomb form by means of heating above 80°C. (c) Catalytic combustion of condensed VOCs with noble metal as catalysator at lower temperature.

Content	Technological parameters	Value in detail
Main draught fan	Capacity of air quantity	50,000 m^3/h
	VOCs concentration of organic exhaust gas	≤1000 mg/m^3
	Rated power	55 kW
	Wind pressure	2800 Pa
Activated carbon with honeycomb form	Size	100 mm × 100 mm × 100 mm
	Wall thickness	0.5 mm
	Density	0.42–0.46 g/cm^3
	Cavity density	100/in^2
	Desorption temperature	≤230°C
Catalysator	Size	100 mm × 100 mm × 100 mm
	Wall thickness	0.33–0.39 mm
	Cavity density	200/in^2
	Content of noble metal (Pt/Pd)	600 ± 10% mg/l
	Rate of water absorption	19–24%
	Softening temperature	≥1380°C
Absorption device	Clearance efficiency	≥95%
	Content of activated carbon	15 m^3
	Wind speed	1.0 m/s
	Rated power	13 kW
	Capacity of air quantity	3500 m^3/h
	Wind pressure	4000 Pa
Catalytic combustion device	burning temperature of VOCs	220–250°C
	Rated power	96 kW
	Capacity of air quantity	1800 m^3/h
	Wind pressure	750 Pa

Table 2.
Technological parameters of equipment for absorption-condensation-desorption-catalytic combustion processing.

also be consumed during VOCs clearance operation of organic exhaust gas due to the large rated power of equipment for both thermal oxidization technologies with combustion.

Besides the technology mentioned above for reduction of VOCs emission during paint processing and protection of atmosphere environment quality, there are still other practical solutions to deal with pollution problem of VOCs emission during actual ship and offshore construction. An environmentally friendly coating material without VOCs, such as waterborne peelable paint, is also gradually employed for ship painting process, which is much more effective to reduce the emission amount of VOCs. Since the production technology of paint without VOCs is still not mature for its mass production, which has higher production cost without popular application in shipyard compared to conventional paint. Advanced painting equipment with high adhesion rate (higher than 70%) is employed to reduce VOCs pollution. In order to enhance the adhesion rate of paint, mixing air spraying equipment with advanced painting technology can be applied to achieve painting processing with high quality of certain thickness and evenness of coating, which also has lots of advantages such as high productivity and utilization efficiency of paint

Moreover, optimization of painting processing is also considered. In addition, painting of a ship piece and block is carried out in advance in a workshop or dry-dock with good condition, while VOCs can be well gathered and recycled to reduce the painting in outdoor yard and dock and atmosphere and environment pollution. In particular, shelter scheme during outdoor painting in some advanced shipyard is applied. In detail, the regions with painting in outdoor yard and dock were insulated and covered with special facilities, while emission amount of VOCs can be controlled without arbitrarily emission.

6. Conclusion

With the above investigation, situations of energy consumption and pollution emission for ship industry in China were presented, while their characteristic and influential factors were also examined; solutions for reducing dust particles and VOCs were introduced and applied to conform to the requirement of environment protection. The following conclusions were also obtained:

1. Fabrication procedures during shipbuilding with energy consumption and pollution emission were systematically introduced and summarized.

2. Consumptions of electrical and chemical energy of each fabrication processing were surveyed and examined, which have large difference due to actual working load, requirement of quality, fabrication approach, and utilization efficiency of energy.

3. All pollutants during shipbuilding were considered, and current emission amounts were also presented, which cause serious harm to atmosphere quality, ecological environment, and human health.

4. Dust particles with massive emission and atmosphere pollution were gathered and controlled with the application of advanced fabrication approach and post-treatment equipment.

5. VOCs as the critical pollutant were holistically examined with recycling and removing technologies based on their characteristics during shipbuilding, and some other solutions were also carried out to reduce the emission for environment protection.

Author details

Jiangchao Wang[1,2*], Bitao Liu[3], Zhangjing Bao[3], Wencheng Jiang[3], Zichao Zhuo[1], Langxiong Gan[4] and Yaqing Shu[4]

1 School of Naval Architecture and Ocean Engineering, Huazhong University of Science and Technology, Wuhan, China

2 Collaborative Innovation Center for Advanced Ship and Deep-Sea Exploration, Shanghai, China

3 China Institute of Marine Technology and Economy, Beijing, China

4 School of Navigation, Wuhan University of Technology, Wuhan, China

*Address all correspondence to: wangjiangchaocn@gmail.com

References

[1] Sands P. The United Nations framework convention on climate change. Review of European Community & International Environmental Law. 1992;**1**(3):270-277

[2] Bodansky D. The United Nations framework convention on climate change: a commentary. Yale Journal of International Law. 1993;**18**:451-558

[3] Bodansky D. The Copenhagen climate change conference: a postmortem. American Journal of International Law. 2010;**104**(2):230-240

[4] Bodansky D. The legal character of the Paris Agreement. Review of European, Comparative & International Environmental Law. 2016;**25**(2):142-150

[5] Savaresi A. The Paris Agreement: a new beginning? Journal of Energy & Natural Resources Law. 2016;**34**(1):16-26

[6] Yuan J, Kang J, Yu C, Hu Z. Energy conservation and emissions reduction in China—progress and prospective. Renewable and Sustainable Energy Reviews. 2011;**15**(9):4334-4347

[7] Liu X, Wen Z. Best available techniques and pollution control: a case study on China's thermal power industry. Journal of Cleaner Production. 2012;**23**(1):113-121

[8] Jiang P, Chen Y, Geng Y. Analysis of the co-benefits of climate change mitigation and air pollution reduction in China. Journal of Cleaner Production. 2013;**58**:130-137

[9] Zhang S, Worrell E, Graus W. Co-benefits of energy efficiency improvement and air pollution abatement in the Chinese iron and steel industry. Energy. 2014;**78**:333-345

[10] Zhou K, Yang S, Shen C. Energy conservation and emission reduction of China's electric power industry. Renewable and Sustainable Energy Reviews. 2015;**45**:10-19

[11] Liu Y, Zhou Y, Wu W. Assessing the impact of population, income and technology on energy consumption and industrial pollutant emissions in China. Applied Energy. 2015;**155**:904-917

[12] Gosens J, Kåberger T, Wang Y. China's next renewable energy revolution: goals and mechanisms in the 13th Five Year Plan for energy. Energy Science & Engineering. 2017;**5**(3):141-155

[13] Chang YC, Wang N. Environmental regulations and emissions trading in China. Energy Policy. 2010;**38**(7):3356-3364

[14] Zheng S, Yi H, Li H. The impacts of provincial energy and environmental policies on air pollution control in China. Renewable and Sustainable Energy Reviews. 2015;**49**:386-394

[15] Jin Y, Andersson H, Zhang S. Air pollution control policies in China: a retrospective and prospects. International journal of environmental research and public health. 2016;**13**(12):1219

[16] Feng L, Liao W. Legislation, plans, and policies for prevention and control of air pollution in China: achievements, challenges, and improvements. Journal of Cleaner Production. 2016;**112**:1549-1558

[17] Liu H, Jin X, Wu L. The impact of marine shipping and its DECA

control on air quality in the Pearl River Delta, China. Science of The Total Environment. 2018;**625**:1476-1485

[18] Qiao B, He W, Tian Y. Ship emission reduction effect evaluation of air pollution control. Transportation Research Procedia. 2017;**25**:3610-3622

[19] Eyres DJ. Ship Construction. Elsevier Publishing; 2006

[20] Celebi UB, Vardar N. Investigation of VOC emissions from indoor and outdoor painting processes in shipyards. Atmospheric Environment. 2008;**42**(22):5685-5695

[21] Chung JW, Lee ME. Case study of hazardous air pollutant concentrations in residential areas nearby small and medium scale shipbuilding companies. Journal of Environmental Science International. 2009;**18**(5):517-525

[22] Malherbe L, Mandin C. VOC emissions during outdoor ship painting and health-risk assessment. Atmospheric Environment. 2007;**41**(30):6322-6330

[23] Wang J, Yi B, Zhang C, Zhou H, Shu Y. Experiments of double curvature plate bending with induction heating and processing parameters investigation by computational analysis. Ocean Engineering. 2019;**192**:106596

[24] Radaj D. Heat Effects of Welding: Temperature Field. Residual Stress: Distortion. Springer Science & Business Media; 2012

[25] Khan FI, Ghoshal AK. Removal of volatile organic compounds from polluted air. Journal of Loss Prevention in the Process Industries. 2000;**13**(6):527-545

[26] Everaert K, Baeyens J. Catalytic combustion of volatile organic compounds. Journal of Hazardous Materials. 2004;**109**(1-3):113-139

[27] Kamal MS, Razzak SA, Hossain MM. Catalytic oxidation of volatile organic compounds (VOCs) – a review. Atmospheric Environment. 2016;**140**:117-134

[28] The emission standard of air pollutants for shipbuilding industry, DB31/934-2015

[29] Emission standard for industrial enterprises noise at boundary, GB 3096-2008

[30] Momber AW, Kovacevic R. Principles of Abrasive Water Jet Machining. Springer Science & Business Media; 2012

[31] Integrated emission standard of air pollutants, GB16297-201

Chapter 5

Development of a Methodology for Monitoring the Deposition Process in Gas Metal Arc Welding (GMAW)

Jairo José Muñoz Chávez, Gerardo Antonio Idrobo Pizo, Margareth Nascimento de Souza Lira and Sadek Crisostomo Absi Alfaro

Abstract

Gas metal arc welding (GMAW) processes need to guarantee the quality of the parts produced from this stability in the manufacture of a single bead. In addition to the visual inspection and subsequent characterization with the cutting of the samples, which consists of destructive analysis, it is possible to monitor the entire process and the quality of the part in a non-destructive way. Therefore, this work aims to develop a methodology for the analysis of non-destructive and online weld beads using high-speed cameras, thermal cameras, profilometer and algorithms in MatLab for data processing. The high-speed cameras allow the capture of images of the electric arc and the metallic transfer mode. The thermal cameras, on the other hand, allow the visualization of the melt pool and the temperatures reached during the deposition. Finally, the laser profilometer allows you to make a point cloud of the part and measure its geometry (height, width, height-to-width ratio, wetting angle) online and more accurately than the caliper. With this, it is possible to use the data, both for geometric quality, preliminary parameterization of additive manufacturing, and later for feeding simulations of welding. Promising the monitoring process during manufacturing.

Keywords: monitoring, welding, low-cost, gas metal arc welding, GMAW

1. Introduction

Welding processes with welds presenting defects in their geometry may also present changes in their metallurgical and mechanical characteristics, such as, for example, residual stresses and deformations and heat-affected zones with coarse grains and subsequent cracking. Understanding the variables involved (electrical, geometric, metallurgical) is of great importance to guarantee the final quality of the material.

With regard to the transfer mode, in GMAW, short circuit, globular, spray and pulsed modes can occur [1]. One way to study the metallic transfer mode of a material is through

IntechOpen

the construction of metallic transfer maps, which are schematic graphs that show regions or areas in which a certain type of transfer predominates. The maps serve as a support tool for analysis of the welding process and guidance for carrying out the operation and aim to establish transfer contours in terms of welding variables. Typically the ways constructed are in terms of voltage or arc length versus current or feed speed [2].

The identification of the transfer modes can be done at the instant the welding process takes place or after performing the weld, through the detection and analysis of different signals, generally provided by the arc region. Sound, light, current and voltage electrical signals, the appearance of the arc, spatter, among others, can enable the detection of the transfer mode. Sophisticated techniques, for example, based on analysis of images obtained through high-speed filming, allow obtaining a greater volume of information and determining, with greater security, the metal transfer mode, but they tend to be expensive and time-consuming. One of the most used techniques for the detection and analysis of some transfer modes is the analysis of current and voltage oscillograms, based on the arc current and voltage signals [3]. When it is not intended to see the arc, but the metal deposition, the shadowgraph technique can be used. This is a lighting technique used in conjunction with high-speed cameras in filming metal transfer in GMAW welding processes. The shadowgraph technique makes use of backlit directional illumination with a laser beam as the light source [4].

After shooting, pre-processing is important to improve its characteristics, which depend on the final objective intended for the image. Some processing typically involves techniques for contrast enhancement, noise removal, and region isolation [5]. Other processings used in digital images are erosion, dilation, thresholding, black and white, Gaussian and Fourier filters. It can also perform operations such as smoothing. Equalization and the convolution of matrices or images, for example, the multiplication of the original image with the images that have applied one or more filters. The dilation process consists of obtaining the reflection about its origin and then shifting this reflection from x. The dilation of A by B is then the set of all x displacements. Erosion is the remainder of displacements; dilation and erosion are dual operations with each other and with inverse behavior [6].

Burnoff's criterion [1] or droplet detachment allows the appropriate wire speed and average current to be found for the material. The methodology seeks a balance between the melting rate of the wire and the wire feed speed. This criterion consists of building a functional relationship to represent all possible conditions of the pulse parameters (peak current, base current, peak time and base time) associated with an average current (Im). This relationship includes a workspace that encompasses all possible combinations of parameters and represents a region called the parametric zone.

For these reasons, this work intends to contribute to the monitoring of the deposition process. The first strategy consists of filming with high-speed cameras using the shadowgraph technique with MatLab programming for image processing in order to monitor the metallic transfer mode, number of drops per pulse, size and droplet detachment. The second monitoring stage consists of using a system with an infrared camera to observe the arc length and geometry of the weld pool and the bead. Finally, the third step consists of validating the measurements using the laser profilometer.

2. Materials and methods

The welding process was carried out on a Fronius Trans Plus Synergic 5000 power source. The power source operated in constant voltage mode and the power source was manipulated using an industrial computer via an interface for ROB 5000 robots. PS 130/6 – 5-P from ABB Robotics and programming for one or more beads with the ROB 5000 Interface, RCU 5000 controller and programming done with the Sport S3 Software. The base metal used was low carbon steel AISI 1020 with the aim of developing the methodology with a better cost-effectiveness ratio, with ESAB OK TUBROD 410 NiMo tubular wire and AWS INOX ER 316 wire, and shielding gas with 94% argon and 6% CO_2 and also pure argon. Tests were carried out combining two types of wire and two types of gases, but the methodology carried out can be adapted for any material and shielding gas that is in demand. The initial studies were carried out in pulsed GMAW mode due to the greater stability of the transfer mode. Decreasing the range of working values for parameters such as average current, peak current, base current, cycle or pulse time, wire feed speed, welding speed and inductance, to obtain a weld with good characteristics. The varied parameters are shown in **Table 1**.

Other aspects of the research that should be highlighted are the metallic transfer technique used and the amount of energy and heat input for the metallic transfer. It is also relevant to point out that the result of these data or parameter values was evaluated in different parts of the process using the shadowgraph technique for the analysis of drops, type of transfer and number of drops per pulse, using, for this purpose, a high-resolution camera. CMOS speed captures 1000 frames per second, enough to capture between 5 and 7 images per pulse, depending on the frequency of the pulse and the frequency of the drop. The camera was placed 5° in relation to the bead and its displacement was always carried out in front of the wire in the same position to visualize it since it is necessary to see the drop, its characteristics, its detachment and the transfer mode. The oscillograms were evaluated using programming made in MatLab, an action that helped in the analysis of the stability of the weld, caloric input, energy per pulse during the process and average current. Then, the Burnoff analysis was performed and the empirical analysis based on data collection was used.

The shadowgraph assembly has as its principle the passage of a collimated laser beam through the arc region so that the resulting image represents the shadow of these elements (wire, metal in transfer, weld bead). An optical bandpass interference filter is placed between the arc and the camera, so that only the laser beam and the

Variable parameters	Fixed parameters
Travel speed: 6–12 mm/s	Contact tip distance piece: 15 mm
Wire feed speed: 2–8 m/min	
Pulse frequency: 55–300 Hz	Torch angle: 0°
Peak current: 300–450 A	
Base current: 15–80 A	Electrode diameter: 1.2 mm
Drop separator current: 50–110 A	
Indutance: 0–50%	Stand off: 12 mm
Base time: 40–50% of period	
Peak time: 40–50% of period	Gas flow: 15 L/min
Drop separator pulse time:0–0.8 ms	

Table 1.
Parameters used in the depositions.

respective shadows will appear in the image, suppressing the light produced by the arc that is not in the region of the filter. This arrangement uses a Galileo-type laser beam expander and the expander uses a diverging lens as the beam input and a converging lens as the output, which ideally produces a flat wavefront at the output of the expander, thus having no distortion or magnification in the geometries. of the elements. The scheme of the technique is illustrated in **Figure 1**.

For the storage of data and/or images, such as, for example, current and voltage data, in addition to the images obtained through the profiling technique, an industrial computer from the ADVANTECH brand, reference ICP-622, was used, which contains a PCI Eagle 703S data acquisition board in charge of acquiring the electrical signals of the process, in addition to being responsible for carrying out the communication between the computer and the ROB 5000 interface, which divides its tasks in the communication providing a time of reading and data response with the power source. The acquisition of signals is done through a program developed in the LabVIEW virtual instrumentation software, in the case of current and voltage measurements. The manipulation of the source for shadowgraph has an additional data acquisition board that records 10,000 data per second, in the case of shadowgraph, the program made through the board is also in charge of manipulating the linear displacement table.

The shadowgraph technique was complemented by the synchronization of the images obtained by filming with the current and voltage oscillograms. The method couples the variations of the welding parameters with the images and makes it possible to correlate the information from the two sources. The shadowgraph, together with the analysis of the oscillograms, allowed the monitoring of the transfer mode in order to elaborate the metallic transfer maps. After capturing the images, pre-processing was carried out to improve their characteristics with erosion, dilation, thresholding, black and white filters, Gaussian and fast Fourier transform techniques.

The second part of the monitoring is based on the development of infrared image capture equipment for detecting output variables such as arc length, width and reinforcement, along with image processing and adjustments in data synchronization, size and organization. To carry out measurements from the ultraviolet to the near-infrared spectrum, the SM-240 spectrometer was used. This equipment has a CCD sensor internally composed of four parts: (1) input mechanism with built-in slot, a fiber coupling adapter and a spectrum classification filter; (2) Czerny-Turner cross-array spectrograph, using high-quality focus; (3) linear charge-coupled device (CCD)

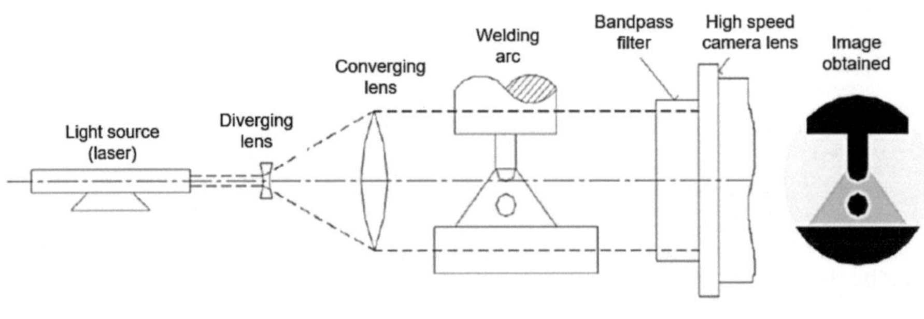

Figure 1.
Scheme used in the shadowgraphy technique [4].

sensor array and driving circuits; (4) Computer interface card. For the measurements, the spectrometer was placed 25 cm from the arc using a 0.2 light attenuator and a tube to direct the light from the weld and reduce the effects of external light. For the adjustment of the camera, a high-power infrared LED with the following characteristics was used:

- LED code: K2005;
- Emission Color: Infrared (IR) 4 chips;
- Power [W]: 5 W;
- Wavelength (nm): 940 nm;
- Luminous Flux or Luminous Intensity: 250 mW;
- Current:700 mA.

The equipment developed to capture the images comprises a 2.0-megapixel Web camera at 30 fps to 50 fps - adapted for IR; lenses, 1000 nm high-pass infrared filter; two 1.0 radiation attenuators; tele objective -zoom focal length (18–108)/Aperture 2.5; and (5); polarizer. To adapt the infrared camera to the camera used for shadowgraph, the location of the infrared camera was perpendicular to the bead and its displacement at 90° and always in front of the wire in the same position to avoid changes in the images in reinforcement and width. The experimental bench with the spectrometer and the two cameras is shown in **Figure 2**.

The laser profilometer was used to validate the geometric measurements performed by the infrared image monitoring system. Calibrations were performed as described in the studies by Idrobo-Pizo et al. [7].

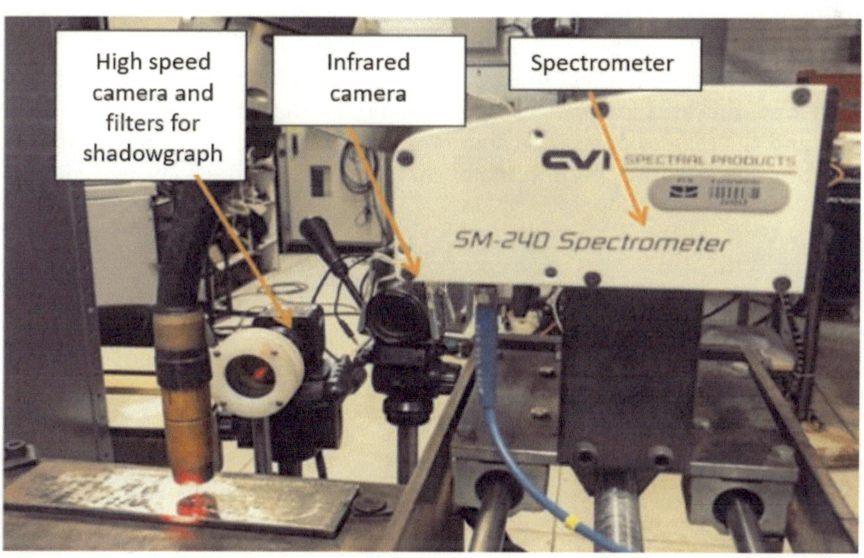

Figure 2.
Experimental bench for making the shadowgraph.

3. Results and discussion

3.1 High-speed camera

As already explained at the beginning of this research, a welding process, to be used successfully, must have the following characteristics: deposit beads free of welding defects; be able to apply the process with a high deposition rate and in all welding positions; absence of spatter and splashes in the formation of deposits; easy opening of the electric arc; produce weld layers with excellent surface finish, among other features.

For the validation of the data collection methods of the developed sensors, it was necessary to legitimize the processes and steps through the shadowgraph technique. For this, it was necessary to carry out the tests synchronized with the capture of images by shadowgraph. Initially, the first method of capturing images by profiling and image pre-processing was used. Tests were performed to capture the droplet mode and characteristics in droplet detachment and arc length performing image processing. **Figure 3** shows one of the tests with a voltage of 26.5 V, wire feed speed of 6.0 m/min and using Argon with 4% CO_2.

The original image has excessive noise making it necessary to treat it. Thus, the second image from left to right received treatment with a Gaussian filter, erosion, dilation and binarization. In the third image, a treatment with erosion, dilation, mean filter and black and white filter was performed. Finally, the fourth image is a superimposition of the treatments carried out in the second and third. Additionally, an algorithm was developed to calculate the relationship between wire diameter, droplet size and arc length. Another analysis is shown below in **Figure 4**.

In the first image, a treatment was carried out to visualize details of the wire and the surface of the pool with a blue background. In the second and third images on the right, different segmentation filters are applied with edge recognition such as Prewitt and Sobel. Finally, the fourth image is created based on the previous two edge recognition images. In this, an algorithm is applied that segments it and recognizes objects by dividing them and placing a different color for each object. The algorithm calculates the number of pixels per area within each object or color and subsequently determines the approximate volume of the drop through a relationship with the calculated area, this area is calculated through a relationship with the pixels within the drop area and the pixels of the wire diameter which is a known value in millimeters. By calculating the maxima and minima of the pixels at the edges, measurements of the

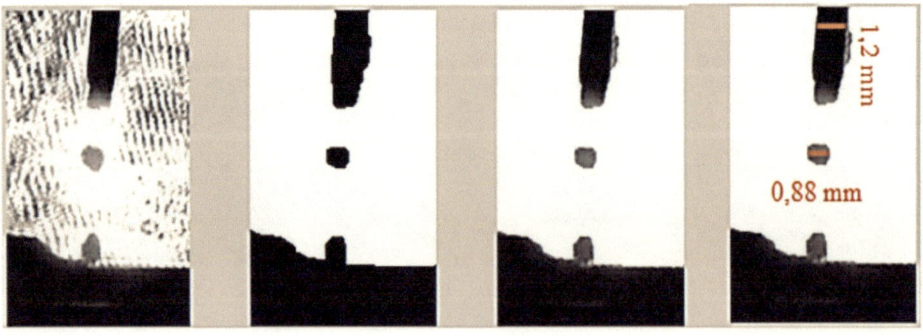

Figure 3.
Shadowgraph image capture for droplet mode.

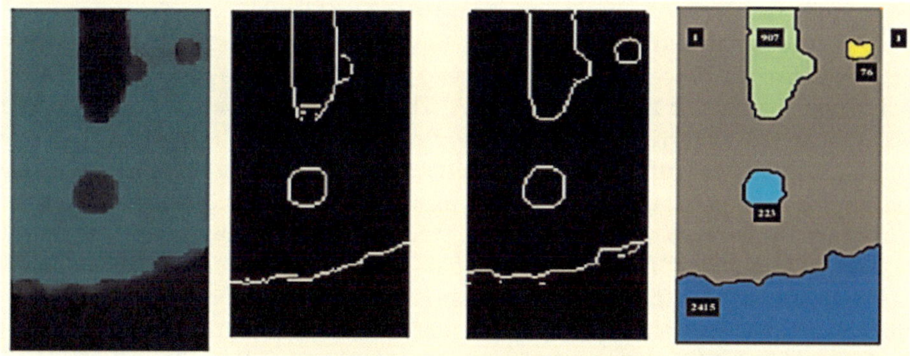

Figure 4.
Edge surface imaging with object recognition and pixel calculation.

arc length can be obtained. By synchronizing these measurements with measurements from other sensors and other variables, it allows for validating measurements from other devices or equipment.

3.1.1 Burnoff criterion

Burnoff's criterion allowed the construction of parametric curves and the study of the regions where the transfer was more stable. Through tests carried out with pulsed current and with constant current, a graph of average current versus wire speed was constructed (**Figure 5**). The intersection point between the two generated lines allowed inferring the range of the transition current between the globular and spray/aerosol transfer modes at 167.8 A with a wire feed speed of 5.04 m/min. Such values can also be decreased or increased in the GMAW-P, using an Im lower than the calculated one, without loss of weld quality. However, if the value of Im is reduced beyond the acceptable level, instability in the process and the formation of weld spatter occur. On the other hand, if the value of Im is increased, the heat generated, the pool penetration and the material flow increase, in addition to converting the

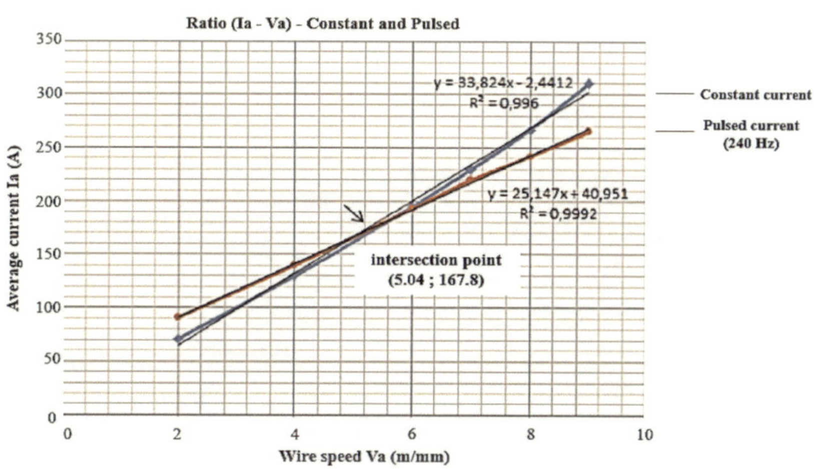

Figure 5.
Relationship between Ia and Va for welding with constant current and pulsed current.

transfer mode from spray to projected spray, bringing instability in the process. In this sense, this study analyzed groups located below and above the intersection point in order to confirm the elucidated aspects. Using wire feed speeds of 5, 6 and 7 m/min with average currents of 175, 188 and 200A. The intersection between the two lines, based on Burnoff's criteria, determines the average current value to obtain a stable spray/aerosol transfer with a droplet diameter size close to the wire diameter and with minimum feed speed for this current. The drop volume is considered equal to that produced by a sphere, as shown in Eq. (1), where r = wire radius (mm) = 0.6 and VD = drop volume (mm^3) = 0.90478.

It is also known that the amount of material deposited on the base metal is the product of the droplet volume and the pulse frequency (F), or the product of the cross-sectional area of the electrode (s) and the velocity (v) Eqs. (2) and (3). With a wire feed speed of 6 m/min, the minimum frequency is F = 125 Hz and the drop formation time, called cycle time, is Tc = 8 ms. Thus, if the drops are smaller and the heat supplied is greater, the frequency must be higher. In this way, drops with smaller dimensions and higher energy have more stability in the metal transfer process. Therefore, the optimal values of time and frequency used in this study had as a criterion the smallest droplet size and the relationships between Im and Va established from the Burnoff criteria.

$$\mathrm{VD} = (4/3).\pi.\mathrm{r}3 \tag{1}$$

$$\mathrm{VD.F} = \mathrm{s.v} \tag{2}$$

$$\mathrm{F} = 3\,\mathrm{v}/(2\mathrm{d}) \tag{3}$$

In this representation, the point of intersection between the two lines was used to calculate the cycle time (Tc = 8 ms) for the studied feed rates, but the values used for this research were Tc = 5 ms and Tc = 6 ms with frequencies of Fr = 166.7 Hz and Fr = 200 Hz, respectively and Im = 180 A in the construction of the Parametric Zones, with the aim of saving energy and increasing stability in the transfer process. **Figure 6** shows the typical time graph of Ip versus Ib for establishing the parametric zone for a feed rate of 6 m/min. Such a representation exhibits a linear relationship between Ip and Ib, the minimum possible value being the common point Im = Ip = Ib. This zone

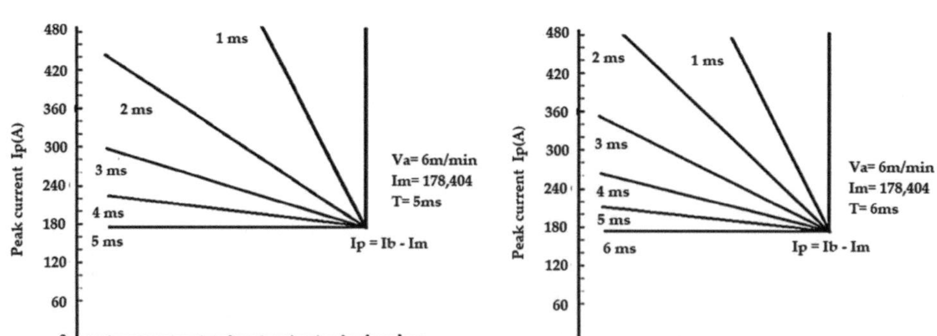

Figure 6.
Parametric zones according to the Burnoff criteria for V = 6 m/min, Im = 178.404A and (a) Tc = 5 ms and (b) Tc = 6 ms.

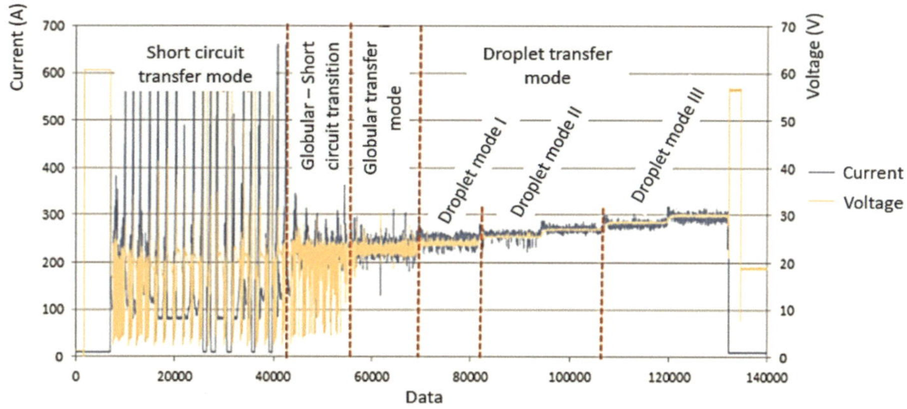

Figure 7.
Voltage scan image to determine transfer modes and process behavior using wire 410 NiMo and 96%Ar + 4%CO_2 gas.

presents the multiple combinations between Ip, Ib and Tp for a given Im, which guarantees a pulse condition that fulfills the Burnoff criterion.

From shadowgraph studies and Burnoff's criteria, correlating with the current and voltage data from the oscillograms, it was possible to draw the current and voltage map showing the regions in which the transfer modes occur, as shown in **Figure 7**. With this, an optimal region of work is chosen depending on the transfer mode and the geometric quality of the bead. As will be further explored below.

3.2 Infrared camera

The results obtained by the spectrometer are shown in **Table 2**. The spectra were captured using a sensor with manual adjustment and calibration of the equipment to obtain the wavelength in meters. For this, LEDs of different wavelengths with known values were used, in addition to considering that the spectrometer has a capture limit range between 100 nm and 900 nm. In addition, the results were crossed with studies by Mota [8] who investigated the spectra between 890 nm and 930 nm, obtaining results similar to those of the present study.

With the data captured by the spectrometer and its subsequent analysis, results were obtained that allowed finding the tracks where it is possible to visualize the bead,

Wavelength (nm)	LED color	Intensity (lm/cm^2) – approx.
320-360-390	Ultraviolet	280–580
460	Blue	150
560–575	Green	200
595	Yellow	200
645	Red	170
815	Infrared	<100
850	High power infrared	<100

Table 2.
Spectrometer results.

opening the possibility of continuing with the project of the CCD sensor of the commercial camera. This sensor can capture a cloud of points of the infrared spectrum forming an image of the bead, in addition, the sensor manages to partially eliminate the spectra generated by the plasma or arc due to its maximum capture limit and the use of filters that restrict the values of capture in the proper range. The images obtained together with image processing and computational algorithms allowed obtaining the necessary information for the research, these being the measurements of width, reinforcement and length of the arc in real time with the 1000 nm filter as shown in **Figure 8**.

One of the objectives of the development of this camera is to obtain the width and the reinforcement using the same camera or the same sensor at the same time. Therefore, if it is necessary to obtain two images in the same camera at the same time, one of the images travels a greater distance as it is an indirect image that passes through two mirrors, which implies obtaining a small blurring of one of the images, which may be the image that arrives directly or the one that reaches the camera indirectly. In the image of **Figure 8**, a better focus can be observed when a single measurement is performed, this is because due to the differences in ways to measure both the width and the length with the same camera, it is necessary to focus the camera on a midpoint between the two paths, which generates a small distortion in the images, a fact that does not occur when only one parameter is being evaluated. To solve this problem, two cameras could be used, but this would increase the cost of the proposed methodology. Another solution comes from the fact that the images can also be improved, using a camera with higher resolution and changing the aperture of the diaphragm to reduce blurring, for this reason, a zoom with variations of the diaphragm and focus was used. Thus, for these tests, a low-resolution 2.0-megapixel web camera was used, which is economical and can capture 33fps or can be adapted to capture 50fps.

One of the tests adopted to carry out the different types of adjustments in the camera, such as, for example, the focus of the two images obtained in the same camera (direct and indirect), the position of the mirrors to capture the indirect image at the bottom of the camera and mainly the test of lighting levels to capture the infrared spectra of wavelengths greater than 950 nm, was the test shown in **Figure 9**, in which three infrared LEDs with different wavelengths (730 nm, 850 nm and 940 nm) were used, as well as high power LEDs each of 5 W. In these tests, it was only possible to capture images of the LED with a wavelength of 940 nm because its transmittance level is 950 $\pm$ 25 nm shown in **Figure 8**, which coincides with the limits of the infrared spectra of the filter used in the camera. The image in **Figure 9** shows the test with the

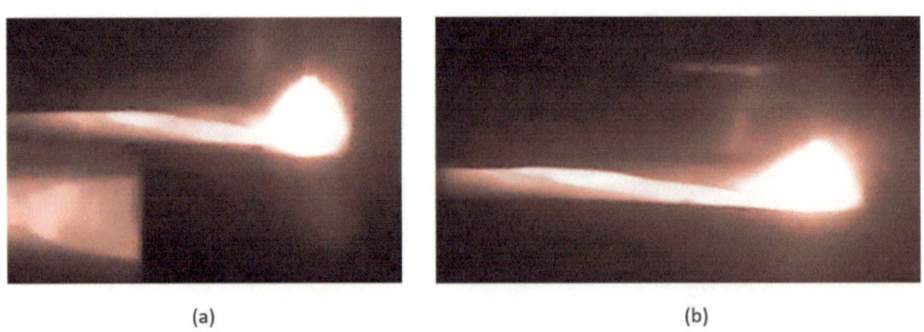

Figure 8.
Measurements tests. (a) Reinforcement and width measurement test in the same camera, using 1000 nm filter, attenuators and polarizer. (b) Test to measure reinforcement only, using 1000 nm filter, attenuators and polarizer.

DOI: http://dx.doi.org/10.5772/intechopen.1002236

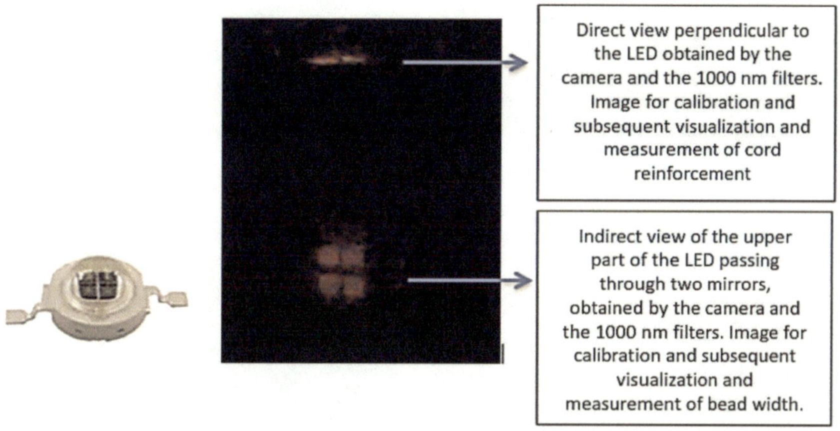

Figure 9.
Infrared camera adjusts with high-power LED 950 nm.

940 nm LED, in the upper part of the image there is a direct image of the profile of the high-power LED, and in the lower part, its front face is seen, captured from a higher position, but with the use of two mirrors taking the image indirectly to the same camera at the bottom of the sensor.

The reason why it is possible to see the 940 nm infrared LED image using the 1000 nm high pass filter is that it can have an emission range with high transmittance up to 975 nm and very low transmittance levels up to 988 nm. Furthermore, the 1000 nm high-pass filter has a tolerance for capturing wavelengths lower than the limit value at −25 nm, so this filter begins to weakly capture the spectra that emit from 975 nm onwards. For this reason, even when testing with a high-power LED, the light intensity captured is weak. As for the other tested LEDs, with shorter wavelengths, it is not possible to capture images.

Figure 10 shows how the algorithm performs the measurements of reinforcement, width and length of the arc with images captured through the infrared camera, the equipment developed and previous treatment of images using media filter, black and white filter and Prewitt edge filter and Sobel.

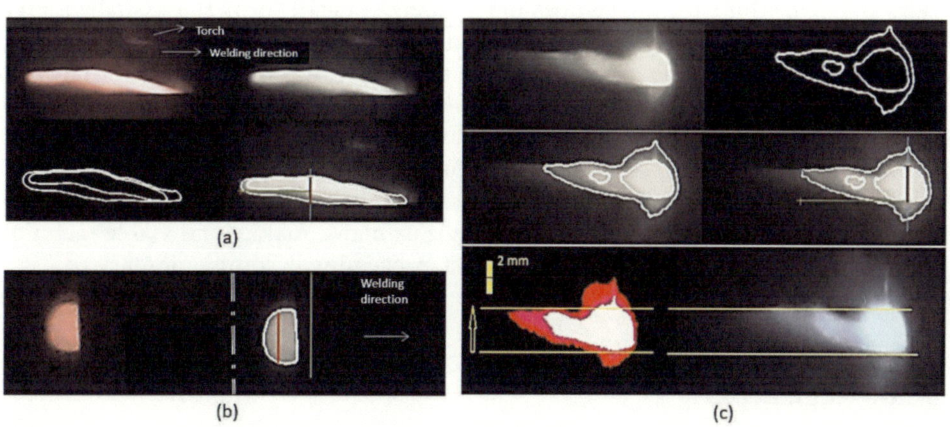

Figure 10.
Measurements with MatLab algorithm of (a) height, (b) width and (c) arch lengh.

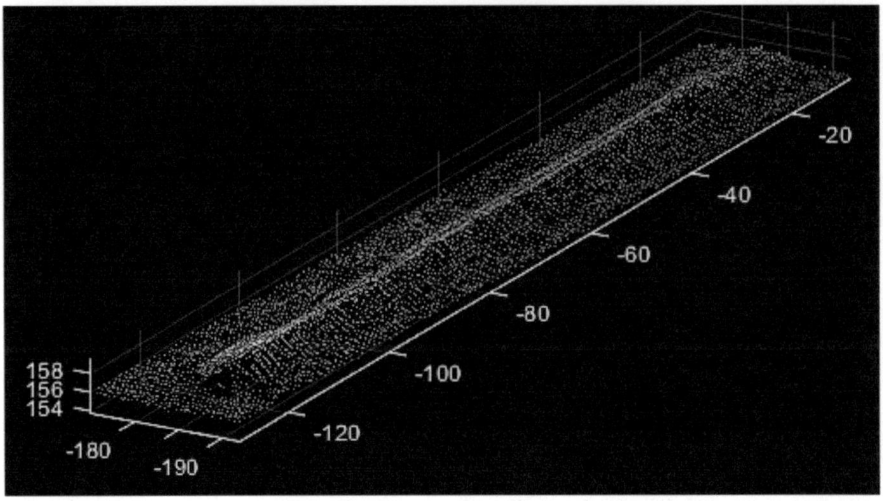

Figure 11.
Point cloud of a bead using a laser profilometer.

3.3 Laser profilometer

To validate the measurements made online with the infrared camera, an offline measurement of the reinforcement and width was carried out, with a scanner for each of the cords obtained, which are presented by a cloud of points in a 3D graph, exemplified in **Figure 11**. The scanner program generates a table with the maximum value data on the Y-axis and X-axis, collected from the point cloud obtained by the laser transversely around the bead. Thus, with this maximum value, determine the reinforcement and width at each measurement instant.

4. Conclusions

It is concluded that high-speed cameras are an effective way to control the transfer mode and parameterize the welding processes together with current and voltage sensors and algorithms for image post-processing, essential for better visualization of the transfer, increasing image quality and reducing noise. All of these techniques, together with the Burnoff criteria, guarantee the choice of the optimal working region and the stability of the welding process and the final quality of the bead. In addition, the use of results obtained via spectrometer and infrared camera allows online obtaining of width, reinforcement and arc length, allowing monitoring of the entire process and bead geometry. Profilometer validation is extremely useful to prove the quality of the monitoring system developed via cameras, making it a viable technique applicable to parameterization processes. Post-processing is complementary to increasing image quality, but online measurements can be performed without using it.

Acknowledgements

To CAPES for financial support and to the University of Brasilia for the necessary infrastructure to carry out the tests.

Conflict of interest

The authors declare no conflict of interest.

Author details

Jairo José Muñoz Chávez[1], Gerardo Antonio Idrobo Pizo[2*],
Margareth Nascimento de Souza Lira[1] and Sadek Crisostomo Absi Alfaro[2]

1 Universidade Federal do Rio de Janeiro, Rio de Janeiro, Brazil

2 Universidade de Brasília, Brasília, Brazil

*Address all correspondence to: gerardo.idrobo@gmail.com

References

[1] Amin M. Pulsed current parameters for arc stability and controlled metal transfer in arc welding. Metals Construction. 1983;**15**:272-278

[2] Scotti A, Ponomarev V. Soldagem MIG/MAG: Melhor Entendimento, Melhor Desempenho. 1st ed. São Paulo: Artliber Editora; 2008. p. 284. DOI: 10.8588098423

[3] Carlson NM, Johnson JA, Smartt HB. Sensing of Metal-Transfer Mode for Process Control of GMAW. In: Thompson DO, Chimenti DE, editors. Review of Progress in Quantitative Nondestructive Evaluation. Boston, MA: Springer; 1990. DOI: 10.1007/978-1-4684-5772-8_252

[4] Balsamo PS, Vilarinho LO, Vilela M, Scotti A. Development of an experimental technique for studying metal transfer in welding: Synchronized Shadowgraphy. In: International Journal for the Joining of Materials. The European Institute for Joining of Materials. 2000;**v 12**:1-12

[5] Gonzales RC, Woods RE. Digital Image Processing. 4th ed. New York, NY: Pearson; 2018. ISBN 9780133356724

[6] Trucco E, Verri A. Introdutory techniques for 3-D computer vision. Computer Science. 1998; CORPUS ID: 44969055

[7] Idrobo-Pizo G, Motta J, Sampaio R. A calibration method for a laser triangulation scanner mounted on a robot arm for surface mapping. Sensors. 2019;**19**:1783. DOI: 10.3390/s19081783

[8] Mota CA, Machado MVR, Fernandes DBF, Vilarinho LO. Study of near-infrared emission on processes of arc welding. Soldagem e Inspeção. 2011;**36** (1):31-39. DOI: 10.1590/S0104-92242011000100006

Chapter 6

Influence of Process Parameters in Gas-Metal Arc Welding (GMAW) of Carbon Steels

Gilbert Tukahirwa and Catherine Wandera

Abstract

Gas-metal arc welding (GMAW) is a widely used fusion joining process in industry known for its ability to create high-quality welds. However, the complex and nonlinear nature of GMAW process variables often leads to weld defects in certain situations. Various dynamic factors affect the GMAW process, including power source dynamics, metal transfer dynamics, arc and droplet dynamics, and weld pool dynamics, making precise control challenging. While achieving completely defect-free welds can be difficult, some welds may still meet industry standards for acceptable performance in specific applications. Properly controlling process parameters plays a crucial role in reducing defects and enhancing overall weld quality. Advanced sensor technologies have emerged as valuable tools for monitoring and adjusting process parameters to achieve desired weld characteristics. This chapter provides a review of research on how GMAW process parameters influence the welding of carbon steels, emphasizing the importance of process control and the role of sensor technologies in improving weld quality.

Keywords: GMAW process parameters, GMAW process dynamics, carbon steel welding, weld defects, process control

1. Introduction

Welding is one of the fundamental processes in the manufacturing industry, which is used to join parts by fusing the workpiece edges to form permanent joints. Numerous welding processes are applied in various manufacturing industries, such as aerospace, automotive, shipbuilding, railway, industrial machinery manufacture, agricultural equipment production, and other general engineering applications [1]. Globally, some of the commonly used conventional welding processes include: shielded metal arc welding (SMAW), oxyfuel gas welding (OFW), gas-metal arc welding (GMAW), flux cored arc welding (FCAW), gas tungsten arc welding (GTAW), submerged arc welding, and spot welding [2]. Moreover, arc welding and gas welding methods like SMAW and OFW are the widely used welding processes in small-size welding enterprises, especially those located in developing countries, because of their cheap cost of acquisition of the welding equipment and low operation costs. However, the SMAW and OFW methods are time-consuming and produce low-quality welds compared to advanced arc welding processes such as the GMAW process that is credited with high-quality welds and flux-free welds [1]. The GMAW process can be used to weld most metals, namely carbon steel, high-alloy and low-alloy steels, stainless steel, aluminum, copper, titanium, and nickel alloys [3].

Due to the continuous wire electrode in GMAW, welds are done at a high speed resulting in time-saving [4]. However, it is inevitable to produce defect-free welds due to the complexity of the GMAW process [5] characterized by nonlinear system variables' relationships [6] and difficulty in setting appropriate input welding parameters—such as welding current, welding voltage, gas flow rate, torch angle, and electrode feed rate—for varying applications [7]. Inappropriate section of process parameters can produce major defects, such as excessive spatter, undercut, overlap, porosity, and poor weld bead contours [1, 8]. Automation with or without robots is an effort toward overcoming defects in GMAW welding [6]. Automation improves both motion and process control resulting in enhanced weld quality; however, inconsistencies in process parameters (inputs) and material variations still hamper the achievement of excellent welding performance in terms of productivity and weld quality [9]. The weld quality is affected by material variations such as thickness, geometry, workpiece fit-up, material composition, and surface finish [9]. For instance, carbon still has a wide spectrum of material compositions ranging from low-carbon steels (mild steels), medium-carbon steels, and high-carbon steels; therefore, selection of the right process parameters for a given composition of carbon steels is of great importance in GMAW welding [10].

Even with the selection of the right material variables and consistent process parameters, the GMAW process still experiences the challenges of too high heat input, too low heat input, arc instability, distortion, and joint position errors due to the nonlinear relationship that exists between variables. The variability in the weld process produces weld defects, such as undercutting, porosity, irregularities in the weld bead, lack of weld penetration, lack of fusion, heat-affected zone (HAZ) softening, microstructure deterioration, crack susceptibility, and loss of mechanical properties in the weld metal (WM) and heat-affected zone [11]. Studies on GMAW welding have developed a considerable understanding on how to address the challenges experienced in the GMAW process in many applications. Some studies have used statistical quality control tools, such as statistical models [12, 13], numerical models [14–18], and artificial intelligent models [19, 20], to predict the desired output parameters by refining the input process parameters [19, 20]. This has led to process advancements in terms of equipment and process control for improvement in the weld quality and productivity, even though some challenges still exist [11].

Artificial Intelligence (AI), one of the Industry 4.0 technologies, can be used to provide intelligent controllers to increase the welding process control [11]. Therefore, the challenge of complex nonlinear relationships of GMAW process variables is expected to be addressed by AI intelligent controllers. This chapter presents the existing knowledge on the influence of process parameters in GMAW welding and the status of GMAW process control for improvement in the weld quality and productivity.

2. Description of gas-metal arc welding

2.1 Gas-metal arc welding process

Gas-metal arc welding (GMAW) is a welding process in which the base metal (workpiece) is melted and subsequently fused by the heat generated from an electric arc struck between the base metal and a continuous consumable bare wire electrode automatically fed from a wire spool at a constant speed through the welding gun; and a shielding gas protects the weld pool from reacting with constituents in the atmosphere [1, 5, 21]. The arc in GMAW has various characteristics that include plasma, temperature, radiation, electrical features, magnetic fields, and arc blow [6]. **Figure 1** shows a schematic illustration of the gas-metal arc welding process.

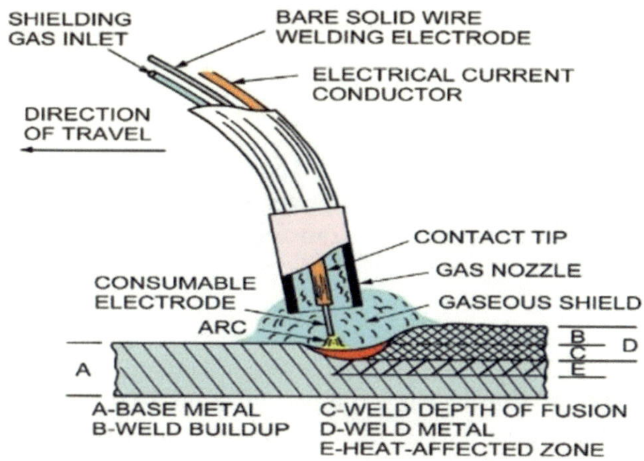

Figure 1.
The gas-metal arc welding process [22].

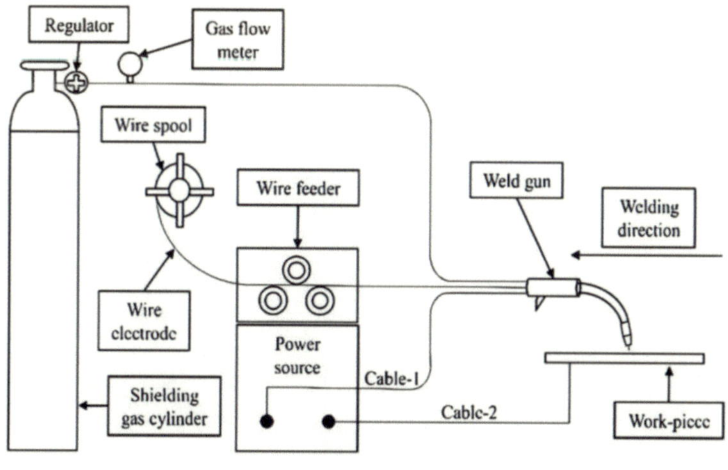

Figure 2.
A schematic layout of the GMAW system components [23].

2.2 Equipment for GMAW

The GMAW process equipment comprises of the electric power source, welding gun, wire electrode feed unit, electrode supply, shielding gas supply, shielding gas hoses, shielding gas regulator and flowmeter, electrode conduit, power and work leads, assorted hand tools, and spare parts, as shown in **Figure 2** [1]. The GMAW process is adaptable to semiautomatic, automatic, automated, and robotic welding [24].

3. GMAW process parameters

The categories of GMAW process parameters presented in **Figure 3** include the power supply variables, shielding gas supply variables, and the workpiece variables that include: voltage, current, speed, filler wire diameter, shielding gas flow rate, the orientation of the electrode, and the distance between the nozzle and plate [6, 19].

3.1 Welding current

Welding current varies directly with wire feed speed or melting rate and an increase in electrode wire diameter necessitates an increase in current [1, 25]. Moreover, welding influences the metal transfer modes [1, 24], penetration [1], and bending strength [4].

3.2 Polarity

Polarity is defined by the electrode that is connected to the electrical terminal; direct current electrode positive (DCEP) or straight polarity is achieved when the wire electrode is connected to the positive terminal, while connecting the wire electrode to the negative terminal gives the direct current electrode negative (DCEN) polarity. DCEP polarity has a stable arc, smooth metal transfer [26], low spatter, good weld bead, and good penetration [5, 27]; consequently, it is the most desirable

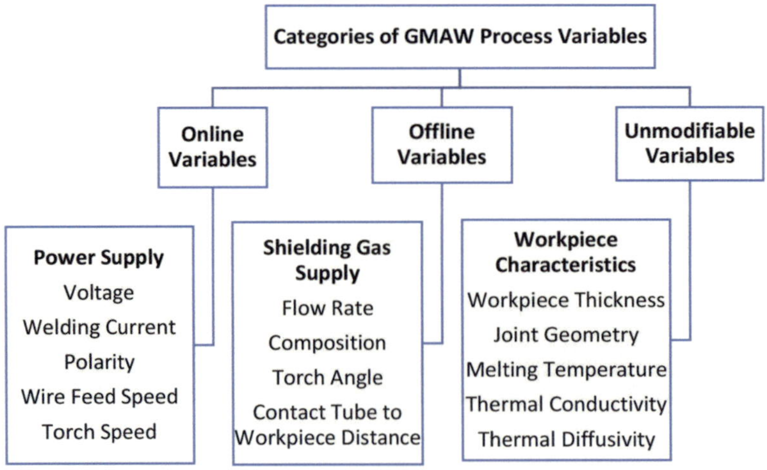

Figure 3.
Categories of GMAW process parameters [6, 19].

polarity [28]. The DCEN polarity produces large drop sizes and arc forces that propel drops away from the workpiece leading to an unstable arc [6]. In the review presented by Kah et al. [29], the alternating current gas-metal arc welding combines the benefit of arc stability of DCEP and high melting rate of DCEN, resulting in high weld quality that is useful in aluminum. Accordingly, the benefits of alternating current-gas metal arc welding (AC-GMAW) process include improved distortion control, reduced heat input, higher welding wire melting rate, and reduced dilution, resulting in higher deposition of filler wire with minimal heat input [29].

3.3 Arc length (arc voltage)

The arc length, which is the gap between the electrode tip and the base of the workpiece base, directly affects the arc voltage; the arc length is inversely proportional to the arc voltage, i.e., a high arc length causes a low arc voltage and vice versa [1, 5, 30]. Additionally, the arc sound pressure increases with an increase in arc voltage [27]. The arc voltage also influences the metal transfer mode [31, 32] and depth of penetration [33].

3.4 Travel speed

Travel speed, which is determined by the rate at which the GMAW torch moves, determines the amount of filler metal that is deposited. The depth of penetration [33] and weld bead geometry [32] are influenced by the travel speed; high filler metal is deposited at slow travel speed, while higher travel speed results in an insufficient filler metal deposition.

3.5 Electrode extension (stickout)

Electrode extension (stickout) is the distance between the contact tube tip and the end of the electrode. An increase in the electrode extension increases resistance causing the end of the electrode to melt before the actual welding, resulting in a reduction in weld heat, penetration, and fusion, and an increased buildup [1, 5].

3.6 Electrode orientation

Electrode orientation is the direction of travel, i.e., trailing and leading orientations. It influences the weld bead geometry and penetration more significantly than the influence of the arc voltage or travel speed on the bed geometry and penetration [5, 20, 30].

3.7 Electrode size

The electrode size influences the transfer mode, deposition rate and determines the amount of current and voltage to be set; a small electrode requires more time to deposit on a thick material, while a larger diameter requires higher current and voltage to be used [1, 5, 34].

3.8 Shielding gases

The main function of the shielding gas in the GMAW process is to protect the molten weld pool from contaminants in the atmosphere and prevent oxidation [23].

Therefore, the shielding gas controls the microstructure and composition of the weldment. The type of shielding gas and its flow rate have a great effect on arc characteristics, metal transfer mode, penetration and weld bead profile, weld speed, etc. A mixture of argon with 2–5% oxygen added, and argon with 25% carbon dioxide gas (CO_2) added or 100% CO_2 (active gas) are used when welding carbon steels [5, 30, 35]. The low gas flow rate can cause blow holes and porosity [17].

4. Weld quality characteristics

The welding quality characteristics that define the GMAW welds include: weld bead geometry (bead width and penetration), heat-affected zone (HAZ), microstructure and mechanical properties (strength, ductility, and hardness) of the weldment), and discontinuities (porosity and spatter) [20, 36, 37]. These weld quality characteristics—which are discussed in detail in the proceeding subsections—are influenced by input (process) parameters such as welding speed, current, wire feed speed, contact tip to workpiece distance, and voltage [1, 5, 19, 24]. The deposition rate is an important process variable that influences the weld quality characteristics.

4.1 Weld bead geometry

The weld bead geometry is characterized by convexity index, depth of penetration (bead penetration), bead height, bead width, and reinforcement area [20, 36, 38]. To obtain the desired quality characteristics, process parameters, such as welding speeds, current, wire feed speed, contact tip to workpiece distance, and voltage, are set on the welding equipment. However, it is hard to maintain these process parameters at a constant value throughout the welding duration due to external disturbances that may cause parameter variations resulting in some undesired quality, such as incomplete penetration, which can contribute to the failure of the welded structure. Proper selection and optimization of the process parameters enable attainment of acceptable weld bead penetration with a high-quality joint.

Different techniques, such as Taguchi experimental design, statistical analyses like analysis of variance (ANOVA), and numerical methods like artificial neural networks (ANNs), have been used to optimize and predict the effect of the process parameters on weld bead geometry [39]. Kalaignar et al. [40] used ANOVA to predict the most significant process parameters affecting weld bead geometry and concluded that welding speed was the major factor affecting the bead penetration and bead width; and that wire feed rate and welding speed were the most significant parameters affecting bead height [40]. A larger penetration is desirable and a smaller convexity index is a desirable quality characteristic [41].

4.2 Heat-affected zone (HAZ)

Heat-affected zone (HAZ) is the part of the metal that changes its properties due to the heat from welding, but does not melt, as shown in **Figure 4** [42]. The features that define the HAZ are the width and grain size, which are influenced by the process parameters set on the welding equipment [35]. Moghaddam et al. [43] employed ANOVA on experimental results and determined that the most significant process parameters that influence HAZ were arc voltage and nozzle-to-plate distance. Too high values of arc voltage and nozzle-to-plate distance resulted in a high heat input, which in

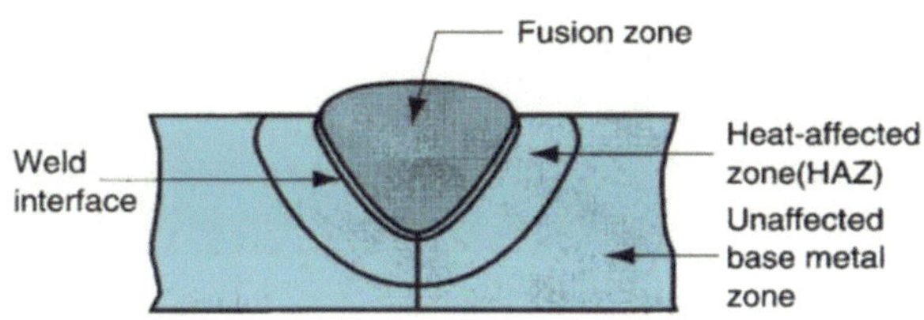

Figure 4.
Heat-affected zone [42].

turn increased the HAZ and vice versa. Therefore, lower values of both arc voltage and nozzle-to-plate distance are desirable for achievement of the desired weld quality [43].

4.3 Microstructure and mechanical properties of the weldment

The heat-affected zone (HAZ) corresponds to the part of the base metal that undergoes structural modifications due to welding. These alterations in the microstructure influence the composition of the welded area and require control, as defects in welding are closely linked to the microstructure of this zone. The microstructure's configuration is interconnected with the mechanical characteristics of the weldment. The arrangement of the microstructure is significantly influenced by the material's chemical composition and the heating and cooling cycles it experiences [44, 45]. Wire feed rate is the most important parameter that affects the HAZ microstructure, as it changes the heat input and cooling rate. Lower wire feed rate leads to smaller HAZ area and ferrite-rich microstructure with some cementite (Fe3C) or austenite. Voltage and welding speed have less influence, and they cause some Widmanstatten Ferrite (WF) and pearlite in the HAZ due to different cooling cycles. The HAZ microstructure is mainly ferrite and pearlite [46].

The HAZ is the part of the base metal that changes its microstructure due to welding. This affects the quality of the weldment and depends on the material's chemistry and the heating and cooling cycles. Wire feed rate is the key parameter that controls the HAZ microstructure of low-carbon steel, as it affects the heat input and cooling rate. Lower wire feed rate makes the HAZ smaller and richer in ferrite with some Fe_3C or austenite. Voltage and welding speed have less effect, and they produce some Widmanstatten Ferrite (WF) and pearlite in the HAZ due to different cooling cycles. The HAZ microstructure is mainly ferrite and pearlite [43, 47].

Figure 5 shows that different microstructures of low-carbon steel are formed under varying welding conditions. The microstructures of similar lead and trail currents and dissimilar currents are clearly visible in the corresponding weld metal (WM) and HAZ. This leads to a corresponding change in the hardness. The size of the grain, its shape (bainitic or polygonal/lath-like or granular), and the composition of phases (bainite/ferrite/pearlite) present determine the hardness in the WM and HAZ. The similarity in hardness values in the HAZ of samples 1 and 2 corroborates the similarity in volume fraction of ferrite grains present. The WM for sample 3 consists of a higher composition of lath-like grains (bainitic ferrite) and some laths are very fine with fine white dots (probably bainite) when compared to sample 4 where large lath-like and granular grains (bainitic and polygonal ferrite, respectively) are dominant. The HAZ for sample 4 also shows softening as it consists of granular ferrite with large grains with very little presence of bainite or pearlite, whereas the HAZ for sample 3 consists largely of granular ferrite with smaller grains, which is responsible for greater hardness [47].

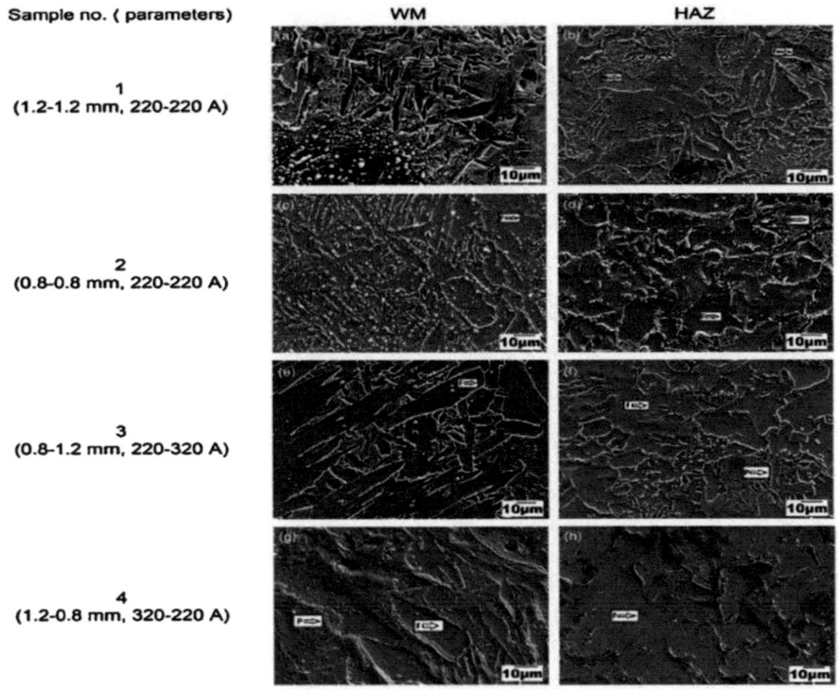

Figure 5.
Microstructures for different wire diameters and welding current [43].

In another experimental study, the effect of changing process parameters, such as current, voltage, root gap, welding rate, and shielding gas flow rate, also showed the change in mechanical properties (ultimate tensile strength (UTS), impact value, root bend, and microstructure) of weldment. The three weldments tested were found to have a microstructure consisting mainly of ferrite grains with a small amount of pearlite. Weldment 2 had finer and more elongated grains than weldment 1 and weldment 3. The specimen of weldment 2 exhibits higher UTS than the other specimens [44]. In a study on the effects of changing varying current, voltage, and number of passes on mechanical properties of HAZ, it was found out that the increase in current increases the hardness, increase in voltage increases the depth of penetration, and increase in the number of passes decreases the hardness [45].

The impact of heat input on the microstructures of two carbon steels with different carbon levels was investigated in a study. For low-carbon steel, there was a decrease in the pearlite with an increase of heat input causing coarser microstructure in the HAZ, while high-carbon steel showed that there was an increase in the pearlite with an increase of heat input causing less coarse microstructure in the HAZ, indicating that high-carbon steel has better mechanical properties and is more resistant to heat input effects [48].

4.4 Discontinuities (porosity and spatter)

According to the American Welding Society (AWS) publication "Standard Welding Terms and Definitions" (AWS A3.0:2001), a "discontinuity" is termed as a disruption in the normal structure of a material which hinders uniformity of its mechanical, metallurgical, or physical properties. In fact, not all discontinuities are considered to be defects. A defect can be a gap in the material that can minimize the accepted standard of a product. A flaw is can be termed as an unwelcome discontinuity. The size and shape of discontinuities vary and bigger size and shapes can minimize weld quality [9].

Porosity, a type of discontinuity in welding, emerges when gas becomes trapped within a weld, leading to the creation of circular or elongated voids, either on the surface or beneath it, as shown in **Figure 6a** [9]. In processes like arc welding, this phenomenon results from gases that are normally dissolved within the molten weld material. If the concentration of these dissolved gases surpasses their capacity to dissolve, the excess gas is expelled from the solution as bubbles or gas pockets during the solidification phase of the weld material [49]. Some causes of porosity are impurities, insufficient shielding, erratic arc, short arc gap, and low-quality welding skill in general [50].

Surface irregularities are flaws on the weld surface that may affect the weld quality and performance. They include ripples, spatters, craters, arc strikes, and pores. Spatter denotes the presence of melted metal fragments that do not properly blend into the weld, possibly indicating problems with the welding procedure or insufficient skill [9]. This issue encompasses the formation of liquid droplets close to the welding arc while welding and their subsequent expulsion from the weld pool, as shown in **Figure 6b**. Multiple factors can trigger this expulsion of melted material, including issues like unsuccessful arc initiation. Insufficiently fine-tuned settings frequently play a role in causing spatter. The consequences of spatter can encompass substandard weld quality, disorderly workspaces, and wastage of materials. Moreover, the removal of spatter from the surfaces of workpieces prolongs production time and raises costs [49].

Sefton and James [51] carried out a research that focused on the challenge posed by GMA welding's sensitivity to strong air currents. They explored the consequences of altering air velocities and shielding gas rates during welding while maintaining consistent other conditions. A specialized apparatus ensured precise airflow control. At draught speeds as modest as 0.9 m/s, the shielding gas dispersed and polluted the weld, resulting in the emergence of porosity, a notable defect. Elevated gas flows without air disruption prevented porosity. Amplified draught speeds and particular gas flows corresponded to heightened porosity, reducing the weld's strength. The study also observed spatter, with porosity as the primary influencer of spatter levels. In fact, the higher the porosity, the higher the spatter and vice versa [51].

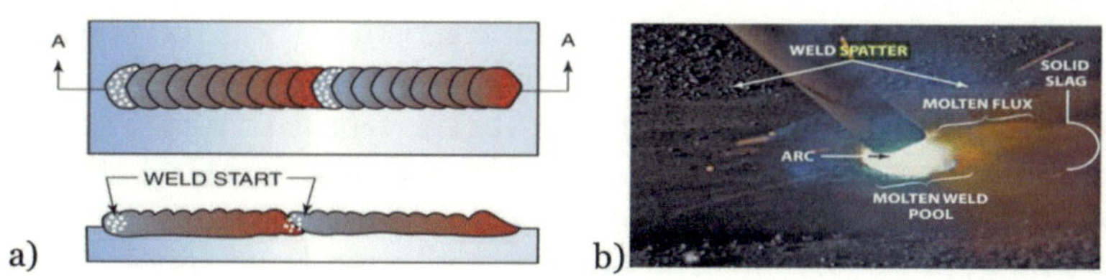

Figure 6.
(a) Clustered porosity and (b) weld spatter [1].

5. Characteristics of carbon steel

Carbon steel is among the commercially important metals with intrinsic mechanical properties, low cost, abundant sources, and high melting temperatures that lead to their intensive and extensive use in various economic sectors such as engineering, transportation, agriculture, and information technology [3, 48, 52, 53]. Various types of carbon steel include the low-carbon steel that comprises not more than 0.20% C, medium-carbon steel that contains the carbon percentage between 0.20 and 0.50%, and high-carbon steel that has more than 0.50% C [42]. The carbon content and alloying elements added determine the characteristics of carbon steel. These characteristics include hardenability, strength, hardness, toughness, wear resistance, workability, weldability, and machinability [54].

5.1 Chemical composition of carbon steel

Chemical composition can be defined as a combination of different elements that make up a chemical substance. A combination of iron (Fe) and carbon (C) coupled with suitable alloying elements makes up carbon steel. The chemical composition of carbon steels varies with the carbon content where low-, medium-, and high-carbon content corresponds with the respective carbon steel types. The common alloys added to carbon steels or that occur as impurities include manganese, phosphorus, sulfur, silicon, copper, nickel, and chromium, among others. Generally, every alloying element composed in the carbon steel material has either negative or positive contributions to its inherent properties and characteristics that include hardenability, strength, hardness, toughness, wear resistance, workability, weldability, and machinability. In fact, the higher the alloying content, the greater the effects it imparts on carbon steel. The chemical composition in the weld metal is basically the combination of base metal and electrode metal compositions. For example, high-carbon content increases the strength, hardness, and wear resistance while it reduces the ductility, weldability, and toughness of the carbon steels [54]. The chemical composition of the weld interface and heat-affected zone is similar to that of the base metal. However, the mechanical properties may differ depending on the cooling rates and temperatures expressed during the welding process [42].

5.2 Welding characteristics of carbon steel

All carbon steels can be welded with low-carbon steel exhibiting a high rate of weldability due to the low-carbon equivalent value which disregards the need for preheat and postheat treatment controls as in the case of high-carbon steel [51, 52]. This chapter examines previous studies and argues that it is important to distinguish between weld characteristics and welding characteristics. These terms may seem similar, but this analogy is not true. Weld characteristics are the properties of a carbon steel-welded joint such as mechanical properties like tensile strength, hardness, elongation, and impact energy [55]. Welding characteristics are the behaviors and qualities of carbon steel during the welding process. These welding characteristics encompass a range of factors that describe how well a welding process performs and how effectively it joins materials. These factors include the welding process parameters, preheat temperature, interpass temperature, and postweld heat treatment, carbon content, cooling rate, joint design, filler material, base metal conditions, alloying elements, and heat input [24, 56–58]. According to B. Choudhury and

Chandrasekaran [58], the weld bead geometry (shape and size of the weld bead) and the strength (force it can withstand) are examples of welding characteristics that measure weld quality [58].

6. Fundamental principle of the GMAW process control

6.1 GMAW process dynamics

Gas-metal arc welding process dynamics refer to the aspect of rapid changes in the GMAW process variables during the welding process which influences the metal transfer dynamics. The GMAW process experiences dynamic changes in the different components of the GMAW system, which include power supply, wire electrode feed, shielding gas supply, and power and work leads [1]. These components of the GMAW system contribute to the disturbances and uncertainties in the GMAW process [59]. The welding process should be stable to give desired weld quality; however, stability of the welding process is not possible due to disturbances and uncertainties over which the welder has no control, resulting in weld defects like spatter, porosity, excessive light, and fumes [60].

To develop mechanisms for overcoming the disturbances and uncertainties observed during GMAW process, several researchers have developed mathematical (statistical and numerical) models that describe the physical GMAW process by employing fundamental principles of science, statistical and/or experimental techniques [6, 12, 13, 32, 40, 41], and artificial intelligence models [14, 15, 59]. The major aim of these models is to provide a framework for describing and understanding the behavior of the GMAW system in order to provide an effective feedback control [6]. Dynamic models developed have either focused on the specific dynamic behavior of the power source [61], metal transfer [62, 63], arc and droplet [64], and weld pool dynamics [65]. The dynamic models have focused on adjustment of the process parameters to attain the desired heat input and mass flow to the molten pool, which directly influences the weld geometry and greatly improves the weld quality [13].

6.2 GMAW process control

Process control of robotic or automatic equipment involves regulating the sequence of operations, process variables, multiple schedules, and motion in order to produce the desired response [9]. The most common process controllers used in the GMAW process include: Proportional-Integral-Derivative (PID), Integral Sliding Mode Controller (SMC), Model Predictive Control (MPC), Adaptive Control, and Intelligent Control.

Proportional-Integral-Derivative (PID): The PID controller uses a feedback control algorithm to regulate three different parameters which are proportional, integral, and derivative based on the past, present, and future errors between the real output taken from the welding process and the expected output (i.e., the set point) [34, 59]. The PID controller has been demonstrated to have the ability to maintain a constant wire feed speed for an automated GMAW welding bench [61, 62, 66]. By controlling an input wire feed, the output arc current becomes constant, hence imparting the desired quality weld. Furthermore, the PID was also used to control the wire feed speed in a standalone photovoltaic (PV) module wire feeder system (PV-WFS) of the GMAW PV-run equipment [67]. PID controllers are widely used in industrial control systems because of the reduced number of parameters to be tuned.

Integral Sliding Mode Controller (SMC): The integral sliding mode controller controls the welding current and arc voltage of a GMAW system by applying a discontinuous signal [68] to control the dynamics of complex nonlinear systems operating under uncertain conditions [69] such that a desired performance can be attained thus increasing the consistency in weld quality [70]. Moreover, consistency in the weld quality can be achieved by the robust finite-time output tracking of the GMAW system [71].

Model predictive control (MPC): MPC is an advanced effective control algorithm used in industrial applications to predict or estimate the future output of a system from a set output by employing the dynamic model [64, 72]. The MPC is robust in modeling uncertainty and has the capability of dealing with system constraints [73]. A nonlinear model predictive controller has been used to control or regulate droplet detachment frequency without measuring it directly; the welding current and arc length are successfully controlled by the MPC controller [74].

Adaptive control: The adaptive controller has the ability to self-regulate or self-adjust the parameter variations and disturbances in the process [6]. The adaptive control system constantly analyzes the feedback from the sensor and adjusts input parameters as well as learns new patterns during the welding process [72]. The use of sensors allows real-time control and adjustment of process parameters such as welding current and voltage [75], leading to real-time monitoring of the weld bead geometry and real-time evaluation of the weld defects [36]. AI concepts, such as expert control, genetic algorithm, fuzzy logic control, and neural networks, can also be used for developing adaptive control [37]. The combination of artificial neural network (ANN) decision-making software and machine vision system leads to the development of an adaptive (AI)-based GMAW parameter control system [37]. Gyasi et al. [38] developed an adaptive intelligent system using infrared thermography (IRT)-based device and AI system.

Intelligent control: The intelligent control is achieved through the use of neural networks, fuzzy logic, pattern recognition, expert systems, AI, and/or knowledge-based systems. An advantage of techniques like neural networks and fuzzy logic is that they do not require precise mathematical modeling of the welding process, which is a stumbling block for all control techniques to be applied to welding processes [6]. According to Thompson and Alfaro [76], intelligent control is a heuristic, autonomous, nonlinear, and adaptive (with learning) controller. Information theory, computer science, and artificially intelligent can be applied to obtain good results in complex systems modeling. Key intelligent technical elements, which mimic the welder or human sensory organs, include: sensing the welding process for the welder's sense organ function, knowledge extraction and modeling of the welding process for the welder's experience reasoning function, and intelligent control of the welding process for the welder's decision-making operation function. In the era of Industry 4.0 that enables the realization of smart manufacturing including smarter robots, the new techniques and applications of data analysis, learning models, and intelligent control are also important in the control of intelligent welding systems [5, 76].

7. Weld quality measurement systems

7.1 Online weld quality systems

The GMAW process characteristics—such as extreme brightness of the arc radiation, spatters/splash of the hot metal, fumes/smokes, high temperatures—inhibit the welder (operator) from clearly viewing the molten weld pool dynamics and the formed weld bead during the welding process. Therefore, online monitoring is used to continuously compare the actual and desired weld quality values so as to adjust the welding parameters based on such real-time sensing and feedback control [77]. An online weld quality measurement system is the quality measurement inspection done during the welding process. A sensor mounted on the robot head scans the welded surface and sends the analyzed measured data to the welder (operator) or control algorithm, as shown in **Figure 7** [78]; and subsequently, adjustments in the input process parameters can be made. The benefits of using an online weld quality measurement system include higher productivity, lower costs due to limited final inspection, less audit, less repair, and greater reliability of the welded components in automated welding operations, which subsequently guarantees good weld quality [79, 80]. Online weld quality systems have been used to satisfactorily measure weld bead dimensions (groove width, weld bead width, filling depth, and reinforcement height) and detect weld defects (plate displacement, weld bead misalignment, and undercut) [81].

Intelligent models, analytical and simulation techniques, such as soft-computing tools, finite element methods, and thermal models for online monitoring, have been developed to improve the online monitoring of the weld quality [27]. Online monitoring can also be obtained using data mining algorithms that detect the process disturbance with the statistical data [82]. Advanced (i.e., reliable, noncontact, and nondestructive) online sensors used to [80] extract data from the GMAW welding process include fuzzy-logic-based optical sensors, laser profile sensors [78], and the hall effect sensor [82]. The usability of the online sensors' quality assessment and diagnosis is limited by their speed, time of evaluation (delayed evaluation or post-process evaluation), cost, robustness, and accessibility. Arc sensing provides more flexible online sensing but is also limited by directly measuring only the input and the intermediate variables of the process without detecting the output variables, so it does not give the information about the weld quality. Consequently, Ayoade and Steele [5] experimentally investigated both arc sensing data collection for in-process and sensor technology sensing for postprocess data inspection to establish the method that provided better weld quality [5]. Online surface measurement is also important for determining weld quality in Wire Arc Additive Manufacturing (WAAM) using the

Figure 7.
Real-time scanning using a laser scanner mounted on the robot head [78].

GMAW process, whereby online surface measurement is used to measure each layer of metal added in order to ensure that the desired bead geometry is obtained before the next layer is laid; the bead geometry determines the quality of the next layer [83].

7.2 Offline weld quality systems

Offline weld quality measurement is a postwelding inspection or nondestructive testing (NDT) technique for identifying defects in which the weld bead is measured after welding or away from the welding station to check whether the desired quality has been achieved [82]. Offline weld quality measurement is time-consuming and costly [27]. Welded surface beads can be measured using numerous handheld manual gauges that measure the weld toe radius shown in **Figure 8** [78]. Vision systems are other tools that are used in weld quality measurement by scanning the welded surface and a computerized environment is used for evaluation and that assessment is done by manually placing measurement points (a minimum of 3 points) in the area of interest on the cross-section curve of the measured surface [78].

8. Weld quality standards

The presence of discontinuities and imperfections is a common occurrence in welded structures; therefore, an appropriate standard must be satisfied by a specific weld so that the welded products can be considered to be fit for service in a specific application without failure [24].

8.1 International standard SS-EN ISO 5817

The international standard SS-EN ISO 5817 is a standard applicable to a wide range of welded fabrications, which provides quality levels of imperfection in fusion welded butt and fillet joints (except for beam welding) in all types of steel, nickel, titanium, and their alloys; the SS-EN ISO 5817 is applicable to material thickness ≥ 0.5 mm and covers fully penetrated butt welds and all fillet welds as well as partial-penetrated butt welds. The standard provides three classifications of quality levels designated by symbols B, C, and D, where B is the highest and D has the lowest requirements on the finished welds for each weld discontinuity and imperfection [80, 81]. The standard specifies more than 40 different discontinuities and weld imperfections for the different quality levels [78], with 23 outside imperfections, 13 inside imperfections, two weld geometry imperfections, and two types of multiple imperfections [84]. The SS-EN ISO 5817 standard was established when stress analysis was based

Figure 8.
Manual weld geometry measurement gauges: (a) radius gauge, (b) radius master, and (c) gauge for throat thickness and depth of undercut [78].

on simple strength of materials calculations (nominal methods) and the acceptance limits were set from a production point of view of good workmanship [85]. However, an inconsistency between acceptance limits according to the SS-EN ISO 5817 standard and fatigue life exists within the same quality level where some existing imperfection types lead to short fatigue life, other imperfection types lead to long fatigue life while some other imperfection types do not influence fatigue life [79, 81].

8.2 Volvo group STD 181: 0004

The Volvo Group STD 181–0004 standard was developed to mitigate the inconsistencies of the SS-EN ISO 5817, which featured a weak relationship between the acceptance limits and the influence of existing imperfections on the fatigue life. The Volvo Group STD 181–0004 standard has five quality levels whereby four quality levels (VE, VD, VC, and VB) are for fatigue-loaded structures and one for static-loaded structures (VS) within general lower demands. Quality level VE, which has the lowest requirements regarding discontinuities, is used for welds where the root and weld penetration is critical; VD is used for as-welded normal quality, VC is used for as-welded high quality, and the highest level VB stands for B posttreated welds [81, 84].

According to Jonsson et al. [82], the Volvo Group STD 181–0004 standard achieves a more consistent weld quality compared to SS-EN ISO 5817, according to the following three important principles:

i. The same fatigue life should be attainable for all imperfection types in one weld quality class level.

ii. A shift from one weld class level to the next weld class level reflects a 25% increase in fatigue strength, which is a factor-of-2 increase in fatigue life.

iii. Only imperfection types that influence fatigue strength are defined in the different classes.

8.3 IIW guidelines on weld quality

The International Institute of Welding (IIW) guidelines on weld quality quantitatively relate weld acceptance criteria to the expected structural performance in terms of fatigue strength [85]. The IIW guidelines on weld quality ensure that the welded joint's imperfections meet the fatigue strength demands, which are defined in S-N curves for different fatigue assessment methods. The IIW guidelines on weld quality define acceptance criteria based on weld geometry and imperfections with a focus on increased fatigue strength [78].

9. Conclusion

Significant weld quality improvements have been achieved by GMAW process dynamics modeling and coupled with adaptive controls that refine process parameter inputs. Online and offline weld quality measurements have been combined to create better weld surface geometry measurements, resulting in the adjustment of process parameters for control of the weld quality. The welding process parameters play a critical role in determining the microstructure and mechanical properties of carbon steel. Therefore, to achieve the required microstructure and mechanical properties while minimizing weld defects, the welding process parameters need to be controlled and optimum welding process parameters need to be found. The process parameters have an effect of changing the microstructure and mechanical properties of carbon steel. Moreover, significant advancements have been made in the areas of sensor technologies and computational power for online and offline weld quality measurement systems of the weld surface geometry has enabled the realization of intelligent automated GMAW processes. Consequently, smart manufacturing with smarter robot welding units, more-intelligent process control, and automated quantitative quality inspections within the welding industry has been realized [5]. The IIW guidelines on weld quality provide acceptance criteria for high-quality weld structures, therefore, the desired weld quality characteristics can be achieved. Furthermore, Smart manufacturing, a technology of the Fourth Industrial Revolution (4IR) (also referred to as Industry 4.0), is now beginning to take center stage in the development of industrial processes. Therefore, efforts are being put into developing smarter welding intelligent technologies that are capable of recognizing abnormalities in the weld, identifying the cause of the weld defect, and making immediate necessary corrective actions leading to weld quality improvement without human intervention.

Conflict of interest

The authors declare no conflict of interest.

Author details

Gilbert Tukahirwa* and Catherine Wandera
Faculty of Engineering, Department of Mechanical and Production Engineering, Kyambogo University, Kampala, Uganda

*Address all correspondence to: tukahgilbert2013@gmail.com

References

[1] Jeffus L. Welding and Metal Fabrication. Boston, Massachusetts: Cengage Learning; 2012

[2] Elaru P. Improving project based learning to enhance trainees. In: Skills Acquisition in Welding and Metal Fabrication at Buhimba Technical Institute in Hoima. Kampala, Uganda: Kyambogo University; 2019

[3] Chavda SP, Desai JV, Patel TM, Ksv L. A review on parametric optimization of MIG welding for medium carbon steel using FEA-DOE hybrid Modeling. International Journal of Science and Research Development. 2013;**1**(9):3-6

[4] Pandhare V, Rai P, Lad BK, Das S, Sabiruddin K. Determination of significant factors affecting the bending strength of weld joint prepared by gas metal arc welding. Journal of Mechanical Engineering Research and Technology. 2016;**2**(1):1-12. Available from: http://www.ijmert.net/Current-Issue.php

[5] Ayoade AA, Steele JPH. Welding heterogeneous measurement system for data mining robotic GMAW weld quality. Welding Journal. 2022;**101**:96-110. DOI: 10.29391/2022.101.008

[6] Naidu DS, Dzcelik S, Moore KI. Modeling, Sensing and Control of Gas Metal Arc Welding. 2003

[7] Chitrambalam ST, Ming TW, Mohammad IS, Mat SB. A study on relationship between process variables and weld penetration for gas metal arc welding (GMAW). International Conference and Exhibition on Sustainable Energy and Advanced Materials. 2011:237-244

[8] Jeet S, Sahoo BB, Barua A, Parida B, Kumar Bagal D. A study on relationship between process variables and weld penetration for gas metal arc welding (GMAW). International Journal of Technical Innovation in Modern Engineering & Science. 2018;**4**(8):683-692. Available from: https://www.researchgate.net/publication/327228295

[9] O'Brien ACLJ. Welding Handbook: Welding Science and Technology. Ninth Edit. ed. Vol. 1. Miami: American Welding Society; 2001

[10] Ssempijja D. Investigations into the Mechanical Performance of Ugandan Made Carbon Steel Bars. Kampala, Uganda: Kyambogo University; 2019

[11] Gyasi EA, Handroos H, Kaha P. Survey on artificial intelligence (AI) applied welding: A future scenario of the influence of AI on technological, economic, educational and social changes. In: 29th International Conference on Flexible Automation and Intelligent Manufacturing (FAIM2019). Vol. 38. 2019. pp. 702-714. DOI: 10.1016/j.promfg.2020.01.095

[12] Chandrasekaran RR, Benoit MJ, Barrett JM, Gerlich AP. Multi-variable statistical models for predicting bead geometry in gas metal arc welding travel speed. International Journal of Advanced Manufacturing Technology. 2019;**105**:1573-1584. DOI: 10.1007/s170-019-04355-0

[13] Park JH, Cho DW, Moon HS. CFD simulation of molten pool dynamic behavior on vertical-downward position in P-GMAW process. International Communications in Heat and Mass Transfer. 2022;**132**(105876):1-17. DOI: 10.1016/j.icheatmasstransfer.2021.105876

[14] Hu Z, Hua L, Qin X, Ni M, Ji F, Wu M. Molten pool behaviors and forming appearance of robotic GMAW on complex surface with various welding positions. Journal of Manufacturing Processes. 2021;**64**(January):1359-1376. DOI: 10.1016/j.jmapro.2021.02.061

[15] Rohe M, Stoll BN, Hildebrand J, Reimann J, Bergmann JP. Detecting process anomalies in the gmaw process by acoustic sensing with a convolutional neural network (Cnn) for classification. Journal of Manufacturing Materials Process. 2021;**5**(4):1-17. DOI: 10.3390/jmmp5040135

[16] He Y, Li D, Pan Z, Yu L, Yuan H, Le J. Dynamic modeling of weld bead geometry features in thick plate GMAW based on machine vision, Mdpi. 2020;**20**(7104):1-18. DOI: 10.3390/s20247104

[17] Srivastava S, Garg RK. Process parameter optimization of gas metal arc welding on IS:2062 mild steel using response surface methodology. Journal of Manufacturing Processes. 2017;**25**:296-305. DOI: 10.1016/j.jmapro.2016.12.016

[18] Terner M, Bayarsaikhan TA, Hong HU, Lee JH. Influence of gas metal arc welding parameters on the bead properties in automatic cladding. Journal of Welding and Joining. Feb 2017;**35**(1):16-25. DOI: 10.5781/jwj.2017.35.1.16

[19] Lorza RL, García EM, Rubén FCR, Martínez MÁ. Using Genetic Algorithms with Multi-Objective Optimization to Adjust Finite Element Models of Welded Joints. 2018. DOI: 10.3390/met8040230

[20] Zhao S, Qiu X, Burnett I, Rigby M, Lele A. Statistical characteristics of gas metal arc welding (GMAW) sound. In: Proceedings of the 23rd International Congress on Acoustics. 2019. pp. 7594-7601. DOI: 10.18154/RWTH-CONV-238970

[21] Antonini JM. Health, safety and environmental issues. In: Saleem Hashmi BM, Van Tyne CJ, Batalha GF, editors. Comprehensive Materials Processing. Amsterdam, Netherlands: Elsevier; 2014

[22] Kaputska N. Reciprocating Wire Feed GMAW – An Advanced Short Circuit GMAW Mode. 2014. Available from: https://ewi.org/reciprocating-wire-feed-gmaw-an-advanced-short-circuit-gmaw-mode/EWI

[23] Dinbandhu V, Prajapati JJV, Abhishek K. Advances in gas metal arc welding process: Modifications in short-circuiting transfer mode. In: Davim JP, Gupta K, Gupta K, editors. Handbooks in Advanced Manufacturing, Advanced Welding and Deforming. Amsterdam, Netherlands: Elsevier; 2021. pp. 67-104. DOI: 10.1016/b978-0-12-822049-8.00003-7

[24] Jeffus L. Welding: Principles and Applications. Boston, Massachusetts: Cengage Learning; 2017

[25] Balamurugan S, Senthilkumar B, Kannan T, Surendran P. Influence of welding process parameters on bead geometry - A review. Journal of Mechanical Engineering. 2015;**1**(3). Available from: https://www.researchgate.net/publication/301503796

[26] Pires I, Quintino L, Miranda RM. Analysis of the influence of shielding gas mixtures on the gas metal arc welding metal transfer modes and fume formation rate. Materials and Design. 2007;**28**(5):1623-1631. DOI: 10.1016/j.matdes.2006.02.012

[27] Pal KP, Bhattacharya S, SK. Investigation on arc sound and metal transfer modes for on-line monitoring in pulsed gas metal arc welding related papers, Journal of Materials Processing Technology. 2010;**210**:1397-1410. DOI: 10.1016/j.jmatprotec.2010.03.029

[28] Bazargan-Lari Y, Eghtesad M, Assadsangabi B. Study of internal dynamics stability and regulation of globular-Spray mode of GMAW process via MIMO feedback-linearization scheme. In: 12th International Conference on Intelligent Engineering Systems. 2008. pp. 31-36

[29] Kah P, Suoranta R, Martikainen J. Advanced Gas Metal Arc Welding Processes. 2013. pp. 655-674. DOI: 10.1007/s00170-012-4513-5

[30] Somani CA, Lalwani DI. Experimental investigation of gas metal arc welding (GMAW) process using developed articulator. IOP Conference Series: Materials Science and Engineering. 2018;**455**(012073):1-8. DOI: 10.1088/1757-899X/455/1/012073

[31] Praveen P, Kang MJ, Incheon Y, Korea S. Arc voltage behavior in GMAW-P under different drop transfer modes. Manufacturing Engineering. 2009;**32**(2):196-202

[32] Kumar A, Khurana MK, Yadav PK. Optimization of gas metal arc welding process parameters. IOP Conference Series: Materials Science and Engineering. 2016;**149**(1):1-11. DOI: 10.1088/1757-899X/149/1/012002

[33] Ibrahim IA, Mohamat SA, Amir A, Ghalib A. The effect of gas metal arc welding (GMAW) processes on different welding parameters. Procedia Engineering. 2012;**41**:1502-1506. DOI: 10.1016/j.proeng.2012.07.342

[34] Rao ZH, Hu J, Liao SM, Tsai HL. Modeling of the transport phenomena in GMAW using argon – Helium mixtures. International Journal of Heat and Mass Transfer. 2010;**53**:5707-5721. DOI: 10.1016/j.ijheatmasstransfer.2010.08.009

[35] Adak DK, Mukherjee M, Pal TK. Development of a direct correlation of bead geometry, grain size and HAZ width with the GMAW process parameters on bead-on-plate welds of mild steel. Transactions of the Indian Institute of Metals. 2015;**68**(5):839-849. DOI: 10.1007/s12666-015-0518-8

[36] Mvola B, Kah P, Layus P. Review of current waveform control effects on weld geometry in gas metal arc welding process. International Journal of Advanced Manufacturing Technology. 2018;**96**(9-12):4243-4265. DOI: 10.1007/s00170-018-1879-z

[37] Penttilä S, Kah P, Ratava J, Eskelinen H. Artificial neural network controlled GMAW system: Penetration and quality assurance in a multi-pass butt weld application. International Journal of Advanced Manufacturing Technology. 2019;**105**(7-8):3369-3385. DOI: 10.1007/s00170-019-04424-4

[38] Gyasi EA, Kah P, Penttilä S, Ratava J, Handroos H, Sanbao L. Digitalized automated welding systems for weld quality predictions and reliability. Procedia Manufacturing. 2019;**38**(2019):133-141. DOI: 10.1016/j.promfg.2020.01.018

[39] Sreeraj P, Kannan T, Subhashis M. Simulation and parameter optimization of Gmaw process using neural networks and particle swarm optimization algorithm. International Journal of Mechanical Engineering and Robotics Research. 2013;**2**(1):130-146

[40] Kalaignar R, Rengarajan S, KrishnaMohan P. An investigation and welding characterization of dissimilar joints of alloy steel with SS 410 through. International Journal of Emerging Engineering Research and Technology. 2017;**5**(6):35-40

[41] Sen M, Mukherjee M, Pal TK. Evaluation of correlations between DP- GMAW process parameters and bead geometry. Welding Journal. 2015;**94**(8):265-279

[42] Groover MP. Fundamentals of Modern Manufacturing: Materials, Processes, and Systems. Fourth Edi ed. Hoboken: John Wiley & Sons, Inc.; 2010

[43] Moghaddam MA, Golmezerji R, Kolahan F. Simultaneous optimization of joint edge geometry and process parameters in gas metal arc welding using integrated ANN-PSO approach. Scientia Iranica. 2017;**24**(1):260-273. DOI: 10.24200/sci.2017.4031

[44] Phogat S, Singh R. Experimental study of effect of process parameter of GMAW welding on mechanical properties and microstructure of steel (SAILMA 350 HI). International Journal of Applied Mechanics. 2017;**12**(1):113-123. Available from: http://www.ripublication.com

[45] Sura N, Singh K. Experimental study on effects of process parameters on HAZ of plain carbon steel using GMAW. In: International Conference on Emerging Trends in Science & Technology (ICETST). Vol. 11. 2018. pp. 9-14. Available from: https://www.researchgate.net/publication/325417305

[46] Tawfeek T. Study the influence of gas metal arc welding parameters on the weld metal and heat affected zone microstructures of low carbon steel. International Journal of Engineering & Technology. 2017;**9**(3):2013-2019. DOI: 10.21817/ijet/2017/v9i3/1709030272

[47] Moinuddin SQ, Kapil A, Kohama K, Sharma A, Ito K, Tanaka M. On process-structure-property interconnection in anti-phase synchronised twin-wire GMAW of low carbon steel. Science and Technology of Welding and Joining. 2016;**21**(6):452-459. DOI: 10.1080/13621718.2015.1124960

[48] Syahida N, Nasir M, Khairul M, Abdul A, Ahmad MI. Influence of heat input on carbon steel microstructure. ARPN Journal of Engineering and Applied Sciences. 2017;**12**(8):2689-2697

[49] Ghazvinloo HR, Honarbakhsh-Raouf A, Shadfar N. A comprehensive study on the welded joints appearance in GMAW. Journal of Materials and Environmental Science. 2021;**2021**(10):12. Available from: https://www.researchgate.net/profile/Hamid-Reza-Ghazvinloo/publication/366722992_A_comprehensive_study_on_the_welded_joints_appearance_in_GMAW/links/63b03814097c7832ca7d360c/A-comprehensive-study-on-the-welded-joints-appearance-in-GMAW.pdf

[50] Raj M. Effect of Process Parameters on Mechanical Properties in GMAW. Delhi: Delhi Technological University; 2016

[51] Mamo S, Foden J. Understanding the shielding gas dynamics and improving the weld quality in MIG /MAG welding with respect to draughts. Science, Research and Development. 2013:61-80

[52] Odebiyi OS, Adedayo SM, Tunji LA, Onuorah MO. A review of weldability of carbon steel in arc-based welding processes. Cogent Engineering. 2019;**6**(1):1-32. DOI: 10.1080/23311916.2019.1609180

[53] Prakash SO, Karuppusway P, Gandhi BS. Enhancing the notch tensile strength of GMAW welded AISI 1013 low carbon steel with taguchi optimization. Samriddhi: A Journal Of Physical Sciences, Engineering And Technology. 2021;**13**(01):20-25. DOI: 10.18090/samriddhi.v13i01.5

[54] Kalpakjian S, Schmid SR. Manufacturing Engineering and Technology. Sixth Edit ed. Singapore: Prentice Hall; 2009

[55] Campbell SW, Galloway AM, McPherson NA. Artificial neural network prediction of weld geometry performed using GMAW with alternating shielding gases. Welding Journal. 2012;**91**(6):174s-181s

[56] Khan MS, Bhole SD, Chen DL, Boudreau G, Biro E, Deventer JV. Resistance spot welding characteristics and mechanical properties of galvannealed HSLA 350 steel. Canadian Metallurgical Quarterly. 2009;**48**(3):303-310. DOI: 10.1179/cmq.2009.48.3.303

[57] Chavda SP, Desai JV, Patel TM. A review on optimization of MIG welding parameters using Taguchi's DOE method. International Journal of Engineering & Management Research. 2014;**4**(1):2250-2758

[58] Choudhury B, Chandrasekaran M. Investigation on welding characteristics of aerospace materials - A review. Materials Today: Proceedings. 2017;**4**(8):7519-7526. DOI: 10.1016/j.matpr.2017.07.083

[59] Wong Y, Ling S. Novel classification method of metal transfer modes in gas metal arc welding by real time input electrical impedance. Science and Technology of Welding and Joining. 2014;**19**. DOI: 10.1179/1362171813Y.0000000184

[60] Galeazzi D, Régis HG e S, Pigozzo IO, da Rosa AF, Pereira AS, Marques C. Analysis of current pulse during short-circuit phase in CMT version of GMAW process under a view of additive manufacturing. International Institute of Welding. 2022;**66**:1-13. DOI: 10.1007/s40194-022-01298-1

[61] Dutra JC, Silva RHGE, Bernardi RA, Schwedersky MB, Marques C, Riffel KC. A new interpretative basis for the high performance GMAW process. Soldagem e Inspecao. 2021;**26**(2620):1-8. DOI: 10.1590/0104-9224/SI26.20

[62] Zhang YM, Yang YP, Zhang W, Na SJ. Advanced welding manufacturing: A brief analysis and review of challenges and solutions. Journal of Manufacturing Science and Engineering, Transactions of the ASME. 2020;**142**(11):1-33. DOI: 10.1115/1.4047947

[63] Ribeiro RA, Dos Santos EBF, Assunção PDC, Braga EM, Gerlich AP. Cold wire gas metal arc welding: Droplet transfer and geometry. Welding Journal. 2019;**98**(5):135S-149S. DOI: 10.29391/2019.98.011

[64] Chen SJ, Wang LW, Xiao J, Wei PS. Arc behavior and droplet dynamics of AC GTAW-GMAW hybrid indirect arc. Welding Journal. 2018;**97**(3):91S-98S. DOI: 10.29391/2018.97.008

[65] Mariappan M, Parthasarathi NL, Ravindran R, Lenin K, Raja A. Effect of alternating shielding gases in gas metal arc welding of SA515 gr 70 carbon steel. Materials Research Express. 2021;**8**(9):1-17. DOI: 10.1088/2053-1591/ac21e9

[66] Trigos AE, Garcia-Guarin J, Espinel Blanco EE. Design of a PID control for a prototype of an automated GMAW welding bench. Journal of Physics Conference Series. 2019;**1257**(1):1-8. DOI: 10.1088/1742-6596/1257/1/012001

[67] Hamouda N, Babes B, Boutaghane A, Kahla S, Talbi B. An Enhanced MPPT Method Combining Fractional-Order and Fuzzy Logic PID Controller for a Photovoltaic-Wire Feeder System (PV-WFS). Vol. 174. Heidelberg, Germany: Springer International Publishing; 2021. DOI: 10.1007/978-3-030-63846-7_12

[68] Bera MK, Bandyopadhyay B, Paul AK. Integral sliding mode control for GMAW systems. IFAC Proceedings. 2013;**10**(PART 1):337-342. DOI: 10.3182/20131218-3-IN-2045.00121

[69] Bera MK, Lal Priya PS, Bandyopadhyay B, Paul AK. Discrete-time Sliding Mode Control of GMAW Systems using Infrequent Output Measurements. 2013. DOI: 10.0/Linux-x86_64

[70] Bera MK, Bandyopadhyay B, Paul AK. Robust nonlinear control of GMAW systems-a higher order sliding mode approach. In: Proceedings of the IEEE International Conference on Industrial Technology. 2013. pp. 175-180. DOI: 10.1109/ICIT.2013.6505668

[71] Bera MK, Bandyopadhyay B, Paul AK. Variable gain super-twisting control of GMAW process for pipeline welding. The Journal of Dynamic Systems, Measurement, and Control Transactions of the ASME. 2015;**137**(7): 1-7. DOI: 10.1115/1.4029408

[72] Penttilä S, Kah P, Ratava J, Pirinen M. Penetration and quality control with artificial Neural Network welding system. In: 27th International Ocean and Polar Engineering Conference. San Francisco, United States: ISOPE. 25 June 2017. pp. 54-61

[73] Xia C et al. Model predictive control of layer width in wire arc additive manufacturing. Journal of Manufacturing Processes. 2020;**58**(August):179-186. DOI: 10.1016/j.jmapro.2020.07.060

[74] Sartipizadeh H, Haeri M. Control of droplet detachment frequency in a GMAW process by a hybrid model predictive control. The Journal of Dynamic Systems, Measurement, and Control Transactions of the ASME. 2018;**140**(11):1-10. DOI: 10.1115/1.4040251

[75] Kah P, Shrestha M, Hiltunen E, Martikainen J. Robotic arc welding sensors and programming in industrial applications. International Journal of Mechanical and Materials Engineering. 2015;**10**(1):1-16. DOI: 10.1186/s40712-015-0042-y

[76] Thompson R, Absi Alfaro SC. Intelligent control proposition on gmaw process with machine learning techniques. In: 25th ABCM International Congress of Mechanical Engineering, 2019. 2021. DOI: 10.26678/abcm.cobem2019.cob2019-0263

[77] Cheng Y, Yu R, Zhou Q, Chen H, Yuan W, Zhang Y. Real-time sensing of gas metal arc welding process – A literature review and analysis. Journal of Manufacturing Processes. 2021;**70**:452-469

[78] Stenberg T, Barsoum Z, Åstrand E, Öberg AE, Schneider C, Hedegård J. Quality control and assurance in fabrication of welded structures subjected to fatigue loading. Weld World. 2017;**61**:1003-1015. DOI: 10.1007/s40194-017-0490-5

[79] Wu CS, Polte T, Rehfeldt D. Gas metal arc welding process monitoring and quality evaluation using neural networks. Science and Technology of Welding and Joining. 2000;**5**:324-328. DOI: 10.1179/136217100101538380

[80] Naso D, Turchiano B, Pantaleo P. A fuzzy-logic based optical sensor for online weld defect-detection. IEEE Transactions on Industrial Informatics. 2005;**1**(4):259-273. DOI: 10.1109/TII.2005.857617

[81] Li Y, Li YF, Member S, Wang QL, Xu D, Tan M. Measurement and Defect Detection of the Weld Bead Based on Online Vision Inspection. 2010;**59**(7):1841-1849

[82] Thekkuden DT, Santhakumari A, Sumesh A, Mourad AI, Rameshkumar K. Instant detection of porosity in gas metal arc welding by using probability density distribution and control chart. The International Journal of Advanced Manufacturing Technology. 2018;**95**:4583-4606

[83] Couto MO, Costa RR, Leite AC, Lizarralde F, Rodrigues AG, Payão Filho JC. Weld Bead Width Measurement in a GMAW WAAM System by using Passive Vision. 2020. DOI: 10.48011/asba.v2i1.1121

[84] Bhardwaj S, Ratnayake RMC, Keprate A. Review of weld quality classification standard and post weld fatigue life improvement methods for welded joints. In: Proceedings of 1st International Conference on Structural Damage Modelling and Assessment. 2020;**110**:978-981. DOI: 10.1007/978-981-15-9121-1_20

[85] Jonsson B, Samuelsson J, Marquis GB. Development of weld quality criteria based on fatigue performance. Weld World. 2011;**55**(01112):79-88